Henry A Nicholson

Outlines of natural History for Beginners

Henry A Nicholson

Outlines of natural History for Beginners

ISBN/EAN: 9783743338678

Manufactured in Europe, USA, Canada, Australia, Japa

Cover: Foto ©ninafisch / pixelio.de

Manufactured and distributed by brebook publishing software (www.brebook.com)

Henry A Nicholson

Outlines of natural History for Beginners

CONTENTS.

CHAP.			PAGE
I.	INTRODUCTORY,		1
II.	CLASS RHIZOPODA,		6
III.	" INFUSORIA,		9
IV.	" HYDROZOA,		11
V.	" ACTINOZOA,		14
VI.	" ECHINODERMATA,		17
VII.	" SCOLECIDA,		21
VIII.	" ANNELIDA,		23
IX.	" CRUSTACEA,		27
X.	" ARACHNIDA,		35
XI.	" MYRIAPODA,		39
XII.	" INSECTA,		42
XIII.	" POLYZOA,		47
XIV.	" TUNICATA,		51
XV.	" BRACHIOPODA,		54
XVI.	" LAMELLIBRANCHIATA,		56
XVII.	" GASTEROPODA,		64
XVIII.	" CEPHALOPODA,		71
XIX.	" PISCES,		76
XX.	" AMPHIBIA,		82
XXI.	" REPTILIA,		88
XXII.	" AVES,		93
XXIII.	" MAMMALIA,		99
XXIV.	THE SUB-KINGDOMS,		104
	INDEX,		110

OUTLINES

OF

NATURAL HISTORY.

CHAPTER I.

INTRODUCTORY.

BEFORE commencing the proper subject of this work—namely, a consideration of the leading types of structure as exhibited in the different classes of animals—it may be well briefly to consider how an animal lives. Without entering into any consideration of the much-vexed question whether the substance which composes the bodies of animals is in all cases substantially the same or not, it remains certain that the mere act of living is, in all cases, attended with more or less destruction of the material of which the living body is composed. Every vital act, of whatever kind, is attended by a destruction of the tissue which is concerned in the act. Thus, every movement of the body is effected at the expense of some muscle, and every thought we think is attended by a destruction of a portion of the substance of the brain. Hence comes the notorious fact that no animal can exist without *food*.

If vital actions did not take place at the expense of the already existing substance of the body, then an animal, when once produced, might go on living for an indefinite period, without any necessity of eating. As it is, the process of living is attended with a constant destruction of the living matter of the body, and if this destruction were not counterbalanced in some way, an animal would rapidly waste away and die from the incessant losses of matter, caused by its movements and other vital processes. This result, however, is prevented by the fact that every animal is constantly taking in "food," for the purpose of repairing the losses caused by living, and in this the process of "nutrition" (Latin, *nutrio*, I nourish), essentially consists.

Different animals live upon different kinds of food, but in all cases the food must contain materials which are capable of taking their place in the structure of the animal; otherwise it is not *food* in the proper sense of the term. And, as a matter of fact, the food of all animals, whether it consist of vegetables or of flesh, or of both combined, can be shown to consist of essentially the same constituents, and to be capable of entering into the body of the animal to be nourished. The first condition of nutrition, therefore, is that the animal should be able to get food containing materials which can be built up into its own tissues.

As a general rule, the food which an animal eats cannot be built up *directly* and *without change* into the new structures of the body. On the contrary, the food has to undergo certain changes before it can be employed in making new tissues. These changes are effected by what is commonly called the "digestion" of the food, but what is more properly and in a more general sense known as "assimilation" (Latin, *assimilo*, I make like to). In other words, the food has to be reduced to a common basis having a certain likeness to the tissues which it is intended to replace, before the animal is in a position to make use of it.

In most animals, the process of assimilation is commenced, and to a certain extent carried out, by a more or less complicated digestive apparatus. In any case, the first part of assimilation consists in the melting down of the food into a common nutritive fluid, which contains certain organic compounds which existed in the food to begin with, or were manufactured out of it in the process of digestion. In the higher animals, the food is so acted upon in the stomach and intestines that it forms a complex fluid, which is then "absorbed" or sucked up from the alimentary canal, to form the "blood." The blood, therefore, is to be regarded as an organic fluid which is manufactured out of the food which the animal eats. The blood has dissolved in it the materials which are necessary for making new tissues, and it has to be distributed to the various parts of the body, so that each tissue may take from it the substances which are requisite to repair its losses. This is usually effected by a distinct propulsive organ— the "heart"—which drives the blood to all parts of the body.

What occurs, then, in any of the higher animals, is readily understood. The various parts or tissues of the body are gradually wearing away, and they require fresh material for replacing their waste. This fresh material is contained in the blood, and is derived from the food; and it is incessantly driven to the different parts of the body. As the blood, therefore, circulates through each organ of the body, that organ abstracts from its living current the materials required to restore its losses and its waste.

Not only does each organ of the body take from the blood the materials requisite to repair its losses, but each at the same time throws off into the blood the worn-out and useless materials which have been produced by its slow destruction and wearing away. The result of this is that the blood very rapidly becomes impure, and gets loaded with waste matter as it circulates through the tissues and organs of the body.

It follows from this that the blood has two processes to undergo, if it is to remain in a healthy condition. In the first place, the various tissues of the body are constantly draining away from it nutritive materials, and it has therefore to be constantly receiving a supply of these same materials from the food, as prepared by the digestive organs. In the second place, the various tissues of the body are constantly throwing off into the blood waste matters, and the blood must get rid of these, if it is to retain its purity. This latter process is generally effected by means of distinct breathing organs, in which the blood is exposed to the action of the air. In the air-breathing animals the breathing organs are filled with air directly; in the water-breathing animals the breathing organs are adapted for absorbing the air which is generally dissolved in water. In either case, the waste materials contained in the blood are got rid of by the action of the oxygen contained in the air, which unites with them, and literally *burns* them up.

The main function, then, which any animal has to perform, is to *nourish* itself, or, in other words, to supply itself with materials capable of taking the place of the tissues which become worn out in the discharge of the vital functions. Animals, however, have more to do than this. They have to become acquainted with what is outside themselves, and to hold certain relations with the external world; and these relations come in a general way under the heads of "locomotion" and "sensation." Most animals, namely, can change their relations to external objects, as regards space, by *moving*, and their movements are generally effected by means of special locomotive organs. Most animals possess the power of changing their place bodily, and even those which cannot do this, by reason of their being permanently rooted to one place, can nevertheless move their bodies more or less freely. The organs of locomotion, or rather the *agents* of locomotion, are usually "muscles;" but some animals move about without having any distinct muscles, or indeed any permanent locomotive organs. In most cases, also, the organs of locomotion are brought under the control of the animal's will by means of the special structures which constitute a "nervous system." It is not, however, by any means necessary that a distinct nervous system should be present.

As a general rule, animals can not only change their place, but they are also enabled to obtain some knowledge of the nature and characters of bodies outside themselves, by means of certain "organs of the senses." Thus, the higher animals have special organs for the perception of light, sound, odours, and tastes or flavours, whilst their sense of touch enables them to detect the more material properties of bodies foreign to themselves, as well as of their own bodies. As a general rule, these senses, as in the case of the locomotive organs, are under the control of a "nervous system;" but there are many animals in which a nervous system is absent, and to which we have, nevertheless, no reason to deny the possession of at any rate some of the senses, if in only a rudimentary form.

For the preservation and continued existence of each individual animal it is only necessary that it should be able to nourish itself, and that it should have certain relations with the outer world. Under any circumstances, however, the life of each individual animal comes to an end some time or other, owing to the inevitable failure of the nutritive powers which comes on when the animal has reached a certain age. In other words, death comes sooner or later to each individual animal, in consequence of its reaching a period of its life when it is unable to "assimilate" food enough to replace the constant loss of tissue caused by living. In every kind of animal, therefore, the *individual* dies after a longer or shorter period of existence, but the *kind* or "species" of animal does not die or disappear, because animals are endowed with the power of reproducing their like. Thus, by producing young like themselves, all animals have the power of continuing their kind through successive generations, in spite of the fact that the individuals of each generation perish, each in his appointed time.

It follows from what has been here remarked, that an animal, in its life, discharges three principal sets of functions. It nourishes itself; it has certain relations with the world around it; and it reproduces its like. All animals discharge these three sets of functions, but the higher animals have much more complicated relations with the world outside than the lower. The study of the way in which these functions are discharged by different animals, belongs to the science of Physiology, and will not be discussed here. In what follows, therefore, we shall simply consider the *form* or *structure* of the more important types of animal life.

CHAPTER II.

CLASS RHIZOPODA.

THE animals belonging to this class are mostly very minute, and none of them, except the Sponges, can be said to be familiarly known. Owing to their complexity of structure, comparatively speaking, it will not be advisable to take one of the Sponges as the type of this class. We shall therefore select as an example the little animalcule known as the *Amœba*.

The *Amœba*, or "Proteus-Animalcule" (fig. 1), is an inhabitant of most collections of stagnant water, especially where decaying vegetable matter is present; and it is microscopic in its dimensions. It derives both of its names from its wonderful power of altering its shape (Greek, *amoibos*, changing; Latin, *Proteus*, a sea-god who had the faculty of assuming different shapes). This power it owes to two circumstances. In the first place, the entire body is composed of a soft gelatinous substance, which has very little cohesion, and which can readily flow in different directions. In the second place, the animal has the power of thrusting out thick, blunt, and finger-shaped processes of its body-substance, which may be compared to little roots. Hence the name of the entire class *Rhizopoda* (Greek, *rhiza*, a root ; *podes*, feet).

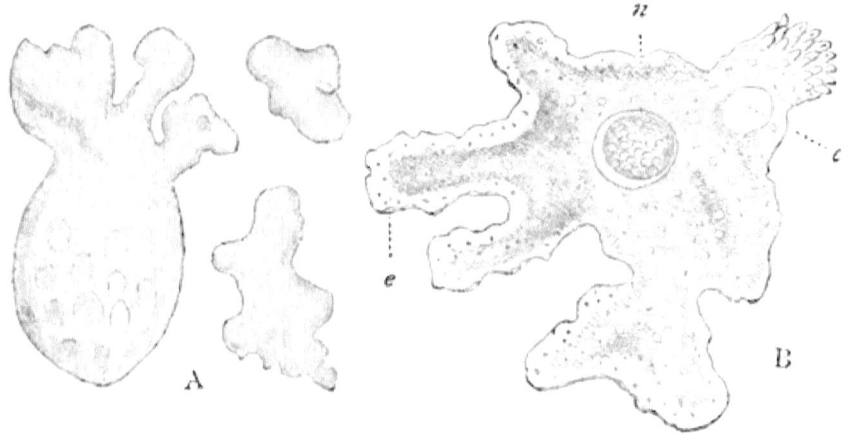

Fig. 1.—A, *Amœba* developed in fluids containing organic matter, very greatly magnified (after Beale). B, *Amœba princeps* (after Carter), highly magnified: c Rudimentary circulatory organ or heart ; e Outer, transparent layer of the body.

These root-like processes can not only be thrust out at any part of the surface of the body, but they can also be withdrawn at will, leaving no traces of their former existence. By means of these, as by temporary feet, the animal creeps about slowly, and by means of these it also obtains its food. Thus, the *Amœba* possesses no mouth, but whenever it comes across any substance or small animal that is eatable, it wraps one of these temporary processes round it, and gradually withdrawing the hand thus extemporised, lodges its capture within the soft substance of its body. Instead, therefore, of possessing a single permanent mouth, the *Amœba* can make a mouth for itself at any point of the surface of the body, and its process of feeding has been appropriately compared to thrusting a stone into a lump of dough.

As before remarked, the substance of the body in the *Amœba* is entirely soft and gelatinous. It exhibits little that can be called "structure," except that the outer layer of the body is more transparent than the central portion, and that this latter is thickly crowded with little rounded particles, and is somewhat more fluid than the exterior. Not only is there no mouth, but there is no trace of any stomach or other digestive organ, and the particles of food, engulfed by the temporary feet, merely pass into the central, semi-fluid substance of the body. There is an organ (fig. 1, B, *n*), which is perhaps concerned with the production of young *Amœbæ*; and there is a little chamber or bladder (*c*), which dilates and contracts at intervals, and which may perhaps be regarded as a rudimentary heart. Beyond these no internal organs of any kind have hitherto been shown to exist.

The *Amœba* is not only to be found in most ponds, but it is usually developed in all fluids which contain vegetable or animal matter, and are allowed to stand exposed to the air in a warm place. Thus, if we take a little hay and boil it in water, and then let the fluid thus prepared stand in any place where the air is allowed free access to it, we shall generally be able to detect *Amœbæ* in it after a few days, by the use of the microscope.

RECAPITULATION OF ESSENTIAL CHARACTERS.—The body is soft and gelatinous, and has no very definite shape. From any part of the surface (or at least from some portion of the surface), processes of the body-substance can be thrust out, which act as temporary feet and hands, and which can be withdrawn when their purpose is fulfilled. There is no mouth, or digestive system, no nervous system, and no breathing-organs; and the circulatory system is at most represented by a rudimentary contractile sac or chamber. These characters distinguish the class of the *Rhizopoda* as a whole.

CHAPTER III.

CLASS INFUSORIA.

THE animals known as "Infusorian Animalcules," are all microscopic in their dimensions, and, though of universal occurrence, they are for this reason only known to scientific observers. They derive their name from the fact that they occur very frequently in what are called "vegetable infusions." By this term is meant the fluid which is obtained by pouring hot water upon any vegetable substance (Latin, *in*, in; *fundo*, I pour), and then straining off the solid particles. The *Infusoria*, then, are so called because they are always, or almost always, to be found in fluids of this nature, which have been exposed for a longer or shorter time to the air. They are also of almost universal occurrence in all stationary collections of water, whether fresh or salt. As the type of this class we may select the animalcule known as *Paramœcium*, a very brief description of which will suffice to indicate the leading points of interest in these animals.

Paramœcium (fig. 2) is one of the most common of animalcules in stagnant water, and is only visible under the microscope. Its body is somewhat oval or slipper-shaped, and is almost quite transparent. Externally it is covered by a delicate membrane, which supports a number of little vibrating hair-like processes, which are called "cilia" (Latin, *cilium*, an eyelash). These little filaments when in active motion cannot be detected by the eye, owing to their transparency, but they can readily be seen when nearly at rest; and they are still more easily recognised by seeing that all minute solid particles which may be floating in the water are forcibly driven away when they approach within a certain distance of the animal. By the vibrations of these "cilia," which lash to and fro like so many little whips, currents are set up in the surrounding water, and particles of food are brought to the mouth of the animal.

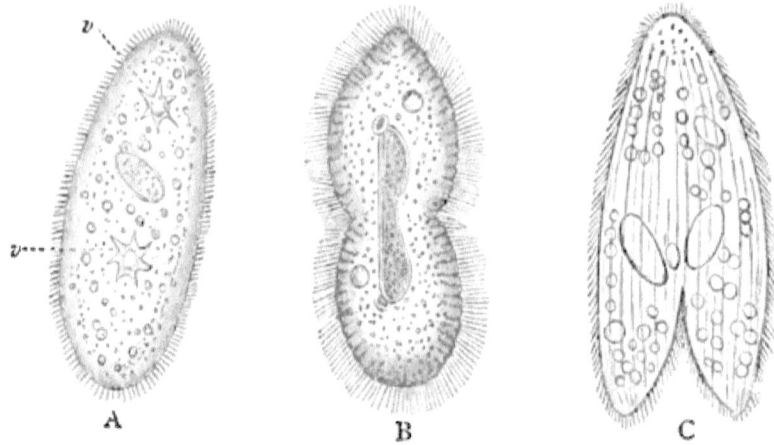

Fig. 2.—A, *Paramœcium*, showing the "cilia" covering the surface, and the rudimentary organs of circulation (*v v*). B, The same, dividing transversely into two halves. C, The same, dividing longitudinally.

Immediately within the delicate external membrane which covers the body, is a layer of a firmer and more consistent character; and this, in turn, gradually melts into a soft, semi-fluid central mass, which constitutes the greater portion of the body, and which contains numerous small solid particles floating in it.

The animal is provided with a distinct mouth, and a short, funnel-shaped gullet, but the mouth does not open into a stomach, or even into a distinct body-cavity. Hence the food simply passes through the mouth into the semi-fluid central substance of the body, and the indigestible parts of it are expelled by a second opening placed near the mouth.

The most important internal organs, and indeed almost the only ones, are one or two little chambers (fig. 2, A, *v v*), which open and shut at regular intervals, perhaps five or six times in a minute, and which appear to drive the fluid which they contain through all parts of the body.

These organs are doubtless a rudimentary form of circulatory apparatus.

There are no distinct organs of digestion, nor breathing organs, nor nervous system, and no traces of a circulatory system, beyond the little contractile chambers just mentioned.

Paramœcium has the power of multiplying itself by dividing into two parts, either transversely or longitudinally (fig. 2, B, C); but it can also produce young by means of eggs.

RECAPITULATION OF ESSENTIAL CHARACTERS.—The body is usually composed of three distinct layers, and is generally provided with the little vibrating filaments which are known as "cilia." A mouth is present, but there is no distinct stomach or body-cavity. There are no digestive or respiratory organs, and no nervous system, and the only traces of a circulatory system are to be found in one or more little pulsating sacs or chambers. These characters distinguish the class *Infusoria* as a whole.

CHAPTER IV.

CLASS HYDROZOA.

THE animals which compose this class are for the most part inhabitants of the sea; and, from their small size, or their plant-like appearance, or again, their living far from land in the open ocean, they can hardly be said to be at all known in general. As the type of this class we shall take a small and abundant form which occurs in many ponds and lakes in Europe, namely, the common Fresh-water Polype (*Hydra vulgaris*), from which the name of the entire class is derived (Greek, *hudra*, a water-dragon, hence a Fresh-water Polype; *zoön*, an animal).

The body of the *Hydra* (fig. 3, A) is in the form of a simple cylindrical tube, the wall of which is composed of two distinct layers, an outer and inner (shown by the dark and light lines in fig. 3, B). At its base the body

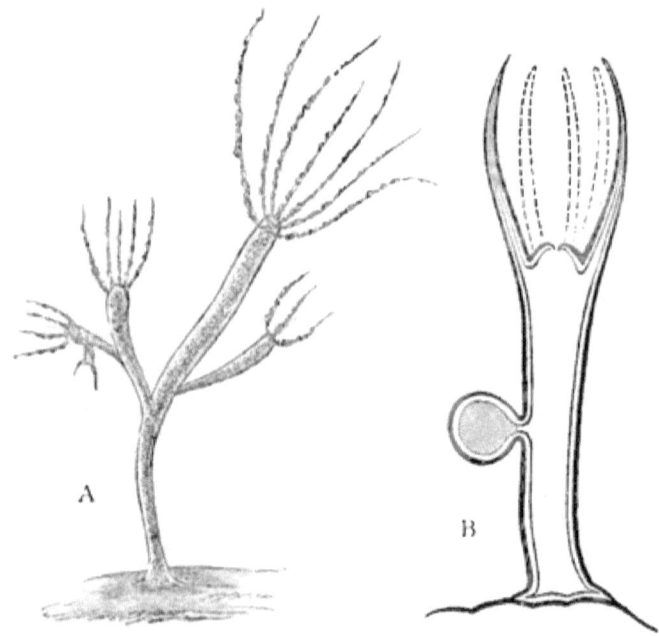

Fig. 3.—A, The Common Hydra (*Hydra vulgaris*), carrying young *Hydræ* which it has produced by budding, considerably magnified (after Hincks); B, Diagrammatic section of the *Hydra*, showing the mouth surrounded by the tentacles, and the disc of attachment; the dark and light lines indicate the two layers of the integument, and on one side of the body is shown a single large egg.

forms a little disc or sucker, by means of which the animal can attach itself to stones, floating pieces of wood, or the stems of water-plants. It can also, by means of this disc, glide slowly, like a snail, over the surface to which it is fixed; but it can detach itself entirely when it chooses.

At the opposite end of the body is placed the opening of the mouth, surrounded by from seven to twelve flexible and extensible processes, which are termed the "tentacles" (Latin, *tento*, I touch). By means of these feelers the animal seizes its prey.

The mouth opens into a long cylindrical cavity, which occupies the whole length of the body, and which receives the food. This cavity may therefore be regarded as the stomach; though it is really the general cavity of the body, since it is enclosed by the general integument. Nothing further can or need be said about the internal anatomy of the *Hydra*; since the undigested food is simply expelled through the mouth, and there are no traces of a nervous system, of breathing-organs, or of any circulatory apparatus.

The walls of the body in the *Hydra* are very soft and contractile, and the animal can pull itself together into a shapeless lump when irritated, or can expand itself and thrust out its tentacles when in search of food. The outer layer of the body, and especially of the tentacles, is roughened by innumerable little microscopic bodies, which are known as "thread-cells" or "nettle-cells." These curious little organs (fig. 4) are seen, when highly magnified, to consist of a little bladder filled with fluid, and carrying at one end a long filament or thread. This thread can be darted out with great rapidity and force, and it is used by the animal in capturing its prey. The thread-cells are too weak to pierce the human skin, but they wound the soft bodies of the worms and other minute animals upon which the *Hydra* feeds, and they appear also to exercise some poisonous or benumbing effect.

Fig. 4.—Thread-cell of the *Hydra*; much magnified.

In the summer time, the *Hydra* produces young ones by a process of budding, as is seen in fig. 3, A, much as a plant throws out buds; but these young are detached to lead a separate life, when they are fully grown. In the autumn the *Hydra* also produces eggs.

The Common Hydra is found abundantly in most

streams and ponds. Though not very much larger than the head of a large pin when contracted, the animal can push itself out to a considerable length, and is generally easily recognised by its brownish-red or orange colour. The *Hydra*, lastly, has a great power of resisting mutilation or mechanical injury. If cut up with a knife, all the pieces will grow and develop themselves into fresh *Hydræ*, and the animal may even be turned inside out, without appearing to suffer thereby.

RECAPITULATION OF ESSENTIAL CHARACTERS.—The body is composed of two distinct membranes or layers, an outer and an inner, of which the outer is furnished with the offensive weapons known as thread-cells. There is a mouth, surrounded by tentacles; but the mouth opens into a large chamber, which may be regarded as the stomach and body-cavity in one. There are no distinct organs of circulation, or respiration, and no traces of a nervous system. These characters distinguish the *Hydrozoa* as a whole.

CHAPTER V.

CLASS ACTINOZOA.

THE chief animals comprised in this class are the so-called Sea-Anemones and Corals; and the scientific name of the class is derived from the fact that the body generally shows a distinctly star-like arrangement of its parts, all of which "radiate" from a common centre (Greek, *aktin*, a ray; *zoön*, an animal). They are therefore "radiated" animals, and a good example of the class may be found in the *Actinia mesembryanthemum*, one of the commonest of the British Sea-anemones.

This familiar Sea-anemone (fig. 5, A) has, when undisturbed and in a state of activity, the form of a short cylinder or column, the base of which forms a broad muscular disc, by means of which the animal fixes itself

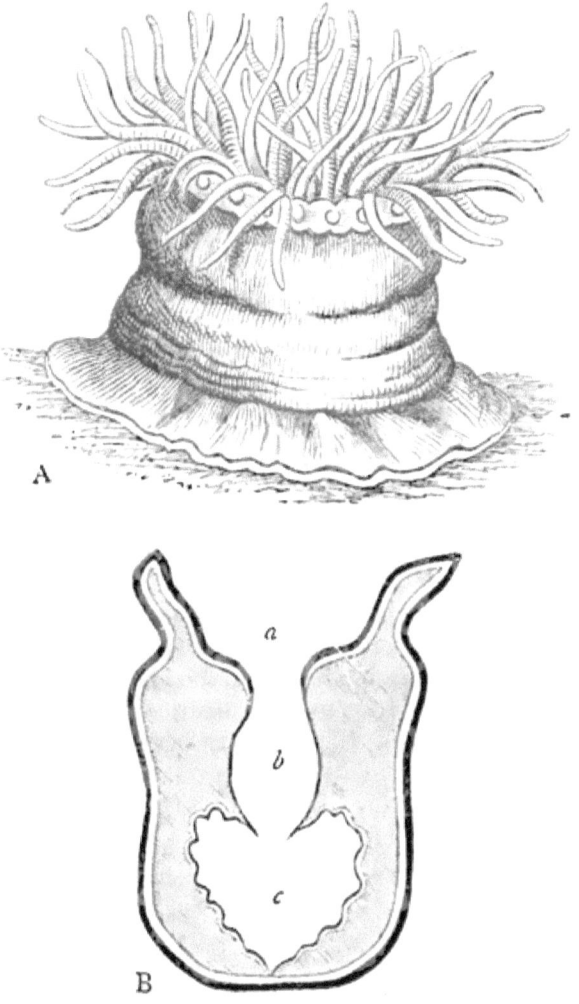

Fig. 5.—A, *Actinia mesembryanthemum*, one of the Sea-anemones (after Johnston); B, Section of the same, showing the mouth (*a*, the stomach (*b*), and the body-cavity (*c*); the dark and light lines show the two integuments of the body.

to a stone or other foreign body. Though thus rooted habitually, the animal has, however, the power of detaching itself and changing its place if necessary. At the end of the animal opposite to the fixed base is placed the mouth, in the centre of a rounded, smooth space, which is surrounded by a fringe of numerous hollow processes or "tentacles."

Both the body and tentacles are formed out of two membranes, an outer and an inner (shown by the dark and light lines in fig. 5, B); and the outer layer of the skin contains a vast number of microscopic stinging organs or "nettle-cells," essentially similar to those which we have already seen in the Fresh-water Polype. The skin, also, is very contractile, and the animal can pull itself together into a mass when irritated, or even when left uncovered by the retreating tide.

The tentacles are hollow, and communicate with the cavity of the body. They can be drawn in at will, and are the organs by which the animal captures its prey. At their bases is seen a circle of from eight to twenty-five bright blue spots, which are perhaps rudimentary eyes.

The internal anatomy of the Sea-anemone is of a very simple character. The mouth opens into a globular stomach (fig. 5, B, b), and this opens below by a wide aperture into the general cavity or space included within the walls of the body. This last-mentioned space is filled with sea-water, mixed with the products of digestion, and it is subdivided by a number of upright membranous plates, which wall off the body-cavity into a series of chambers or compartments.

The digested portions of the food pass through the stomach into the general cavity of the body, and the indigestible portions are got rid of through the mouth. There is no nervous system, nor are there any breathing-organs, nor any organs of the circulation.

The *Actinia mesembryanthemum* is strictly marine, and is found abundantly on the coasts of Britain between tide-marks, adhering to stones, or expanding its beautiful flower-like disc in shallow rock-pools. It varies extremely in colour, being usually of a liver-brown or olive-green colour, but not uncommonly being purplish-red, or grass-green; and it attains a diameter of an inch or an inch and a half.

RECAPITULATION OF ESSENTIAL CHARACTERS.—The body is made up of two distinct layers, enclosing a "body-cavity" which communicates freely with the outer world through the mouth. There intervenes, however, between the mouth and the body-cavity a globular stomach. The integument is furnished with "nettle-cells." There is no nervous system (with few exceptions), and distinct organs of respiration and circulation are not developed. These characters distinguish the class *Actinozoa* as a whole.

CHAPTER VI.

CLASS ECHINODERMATA.

THE commonest of the animals which are included in this class are generally known as Sea-urchins, Star-fishes, Brittle-stars, Sand-stars, and Sea-lilies. Most of these common names refer to the fact that the body in these animals is generally more or less star-like in shape. The name "Sea-urchin" and the technical name "*Echinodermata*," on the other hand, refer to the fact that the skin in these animals is generally rendered prickly, like that of a hedgehog, with numerous spines, tubercles, and grains of lime (Greek, *echinos*, a hedgehog; *derma*, skin). As an example of this class we may take the common Star-fish or Cross-fish (*Uraster rubens*) of British seas.

The most conspicuous point about the form of the Starfish is its strikingly star-like shape (fig. 6). It consists, namely, of a by no means well-marked central body or disc, from which spring five (sometimes four or six) blunt finger-like processes or "arms." The arms, in fact, form a star, and the body looks as if it were composed of the

18   OUTLINES OF NATURAL HISTORY.

bases of the arms united together. The body of the Star-fish is therefore said to have a "radiate" structure (Latin, *radius*, a ray).

Fig. 6.—The Common Star-fish (*Uraster rubens*), natural size, viewed from above.

The entire upper surface of the body in the Star-fish is covered with a leathery skin, from which project numerous blunt conical spines or prickles composed of lime. Along the middle of the back of each arm, these spines form a well-marked zigzag line, more conspicuous in some examples than in others. Between the spines also are much smaller, stalked prickles, which can be seen when magnified to terminate in little pincers.

In the centre of the under surface of the body is placed

the opening of the mouth, surrounded by a fringe of long spines. From the mouth radiate five broad and shallow grooves, which proceed along the under surface of the five arms, gradually tapering towards their extremities. Each of these grooves is bordered by a row of long spines, and each has its floor formed by a double row of little plates (fig. 7, *a a*), which run across the groove, and are so shaped as to leave a little opening between every pair of plates in the series.

It follows from the above, that if we cut the arm of the Star-fish across, we find that it is rounded above, and

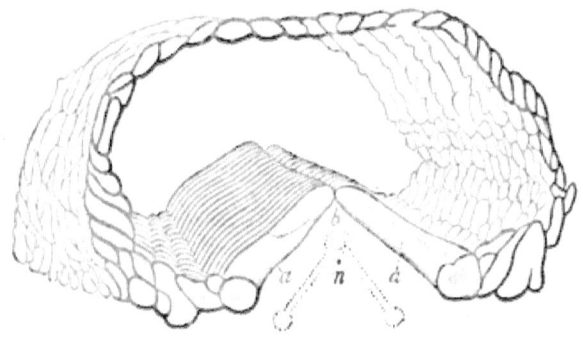

Fig. 7.—One of the arms of the Star-fish, cut across, to show the groove on the under surface. *a a* Transverse plates which form the floor of the groove; *b* The water-vessel, with the little "feet" proceeding from it; *n* Nerve-cord.

hollowed out or grooved below (fig. 7). If we examine the animal in its living state, we find each groove under the arms to be occupied with four rows of little delicate membranous tubes, which end in little suckers. These tubes are called the "feet," because it is by the combined action of these that the Star-fish creeps about. The "feet" can be thrust out to a great length, and they spring from a common tube which runs along the bottom of the groove below the arm. (They are shown in fig. 7 at *b*, but here only two of the rows of feet are shown, the other two being omitted for the sake of clearness).

The tubes which run along the grooves under the arms,

as well as the "feet" proceeding from them, may be called "water-vessels," because they are filled with water from without. The sea-water, in fact, is admitted to them by means of a little grooved, rounded tubercle which is seen on the back of the animal (fig. 6) placed at the angle where two of the arms unite.

The mouth of the Star-fish is not provided with teeth, and opens into a very thin and membranous stomach, which can be thrust forth or "pouted out" through the mouth. From the stomach proceed ten much-branched membranous sacs, two of which are prolonged into each of the arms. The stomach terminates in an intestine, which opens by a minute vent upon the back of the animal.

The nervous system has the form of a circular cord surrounding the mouth, and sending a branch along the groove in each of the arms. At the tip of each arm there is also seen a small reddish spot, surrounded by a circle of spines, and these are of the nature of rudimentary eyes.

There are no distinct breathing-organs, and the process of respiration appears to be chiefly carried on by the absorption of the sea-water through the skin of the back, the delicate membrane which lines the interior of the animal being protruded for this purpose in the form of small tubes which project through interspaces in the integument.

The common Star-fish usually measures from three to six inches across, and is generally of a reddish, yellowish, or orange colour. During life its skin, though very rough and prickly, is comparatively soft; but it contains so much lime that it can be excellently preserved simply by drying it in the sun. It is entirely a native of the sea, and is found from low water to depths of twenty or thirty fathoms. It is very voracious, and feeds upon oysters and other shell-fish, seeming to suck the animal out of the shell by means of the protrusible stomach. Lastly, the Star-fish has the power of reproducing its arms when broken off or injured.

RECAPITULATION OF ESSENTIAL CHARACTERS —The body exhibits more or less clearly a star-like arrangement of its parts. The skin is more or less hardened or roughened by means of lime deposited in it. There is a peculiar system of tubes or "water-vessels," which contain water, usually communicate with the exterior, and are generally employed in locomotion. There is a distinct nervous system consisting of a ring-like cord surrounding the mouth, and sending off branches in a radiating manner. These characters distinguish the class of the *Echinodermata* as a whole.

CHAPTER VII.

CLASS SCOLECIDA.

THE class *Scolecida* (Greek, *skolex*, a worm) includes chiefly the various worm-like animals which live parasitically in the interior of other animals. Besides these there is a number of nearly related forms, which lead a free existence, together with the singular group of the Wheel-animalcules. One of these last we shall select as an example of this class, not as being by any means a *typical* example, but as not presenting certain disadvantages under which other more characteristic forms labour.

If we take one of the free-living Wheel-animalcules, such as *Eosphora* (fig. 8), we find that we have to deal with a microscopic, translucent, little creature, which is to be detected swimming about actively in fresh water. The name of "Wheel-animalcule" refers to one of its most prominent peculiarities—namely, that the front end of the body is in the form of a disc surrounded by a fringe of little vibrating hair-like processes ("cilia"). When these latter are in active motion, the whole head looks as if it were rotating rapidly, like a wheel. By means of this disc the animal not only drives itself through the water, but also sets up currents which bring floating particles of food to the mouth. At the hinder end of the body there is a little pair of pincers, composed of two little diverging "toes," and by means of these the animal can at will moor itself to the stems of aquatic plants. We can also observe that the integument is to a certain extent ringed or transversely wrinkled, though not nearly in such a marked manner as in the true Worms.

As before remarked, the animal is nearly transparent, and its internal anatomy can thus be readily studied. On one side of the locomotive wheel is placed the opening of the mouth (fig. 9, b), to which a depression in the disc conducts. The upper portion of the gullet (c) is much dilated, and contains a complicated series of horny jaws. There is a well-developed stomach (d), and the intestine opens into a chamber, with which the "water-vessels" also communicate.

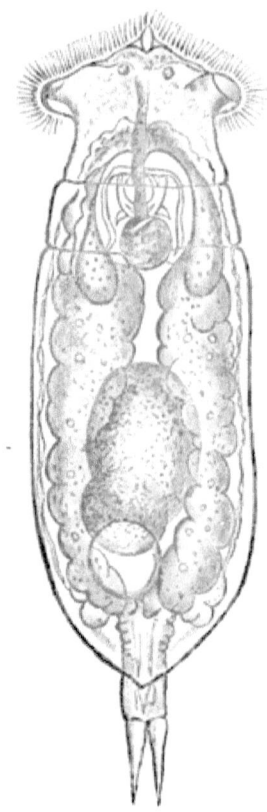

Fig. 3.—Rotifera. *Eosphora aurita*, one of the Wheel-animalcules. Enlarged about 250 diameters. (After Gosse).

The "water-vessels," just alluded to, are two tubes ($g\ g$), which run along the sides of the body, and open behind into a contractile bladder (f). What the exact function of these vessels may be is uncertain, but they are supposed to be connected with the process of respiration.

There is no distinct heart, or true blood-system of vessels. The nervous system, however, is well developed, and consists of a little double nervous mass (h) situated near the head, and carrying upon it the eye, in the form of a brightly-coloured spot.

Wheel-animalcules, such as here described, may be readily detected in most ponds, ditches, or slow-running streams where water-plants grow abundantly. Though microscopic in their dimensions, they are interesting objects of study, owing to the facility with which their transparent skin allows their internal organs to be seen.

RECAPITULATION OF ESSENTIAL CHARACTERS.—The body does not exhibit definite segmentation, nor does it carry upon its sides symmetrically disposed appendages. There is no distinct heart nor blood-vessels, but there is a remarkable set of vessels which usually communicate with the exterior. The nervous system has the form of one or two little nervous masses. These characters distinguish the class of the *Scolecida* as a whole.

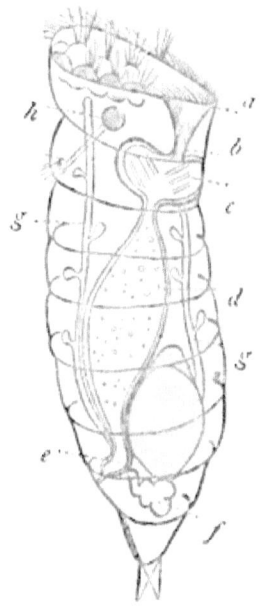

Fig. 9.—Diagram of the anatomy of a Wheel-animalcule. *a* Depression in the wheel-organ leading to the mouth (*b*); *c* Dilated upper portion of the gullet, with the horny jaws; *d* Stomach; *e* Chamber into which the intestine opens; *g g* Water-vessels, opening into contractile chamber (*f*); *h* Nervous system.

CHAPTER VIII.

CLASS ANNELIDA.

THE Ringed Worms, or "Annelides," comprise such animals as the Leeches, Earth-worms, Water-worms, Tubeworms, and Sand-worms, and derive their name from the fact that the integument or skin is markedly ringed or thrown into transverse folds (Latin, *annulus*, a ring). As an example of this class, we may select the Common Medicinal Leech (*Sanguisuga officinalis*).

The Leech (fig. 10, *a*) is of an elongated, cylindrical, worm-like form, tapering towards its head, but capable of greatly contracting or lengthening itself at will. Its skin is quite soft and flexible, slimy to the touch, and very markedly ringed or transversely folded, the total number of rings being about one hundred. At each extremity of the body it shows a little sucking disc or cup (fig. 11, B, *as* and *ps*), by means of which it can move actively about. The hinder sucker (*ps*) is distinctly constricted off from the body by a kind of neck, is not perforated by any aperture, and is provided with an even circular margin. The front sucker is of an elongated form (*as*), is really formed mainly out of the elongated upper lip, and is perforated by the opening of the mouth. When moving, the Leech fixes its hinder sucker, and stretches out its head till it meets with some solid object; it then detaches the hinder sucker, and brings this forward till it adheres close beside the spot where the head has fixed itself. Then, detaching the head, it again extends itself to seek another point of attachment further on. By a repetition of this process, the Leech can travel with considerable rapidity; but it also swims well by a serpentine bending of the body.

The body of the Leech is absolutely destitute of limbs

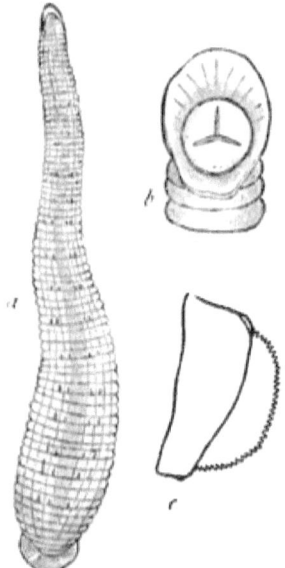

Fig. 10. — *a* The Medicinal Leech (*Sanguisuga officinalis*), natural size; *b* Anterior extremity of the same magnified, showing the sucker and triradiate jaws; *c* One of the jaws detached, showing the semicircular toothed margin.

or appendages of any kind, and locomotion is entirely effected in the manner just alluded to.

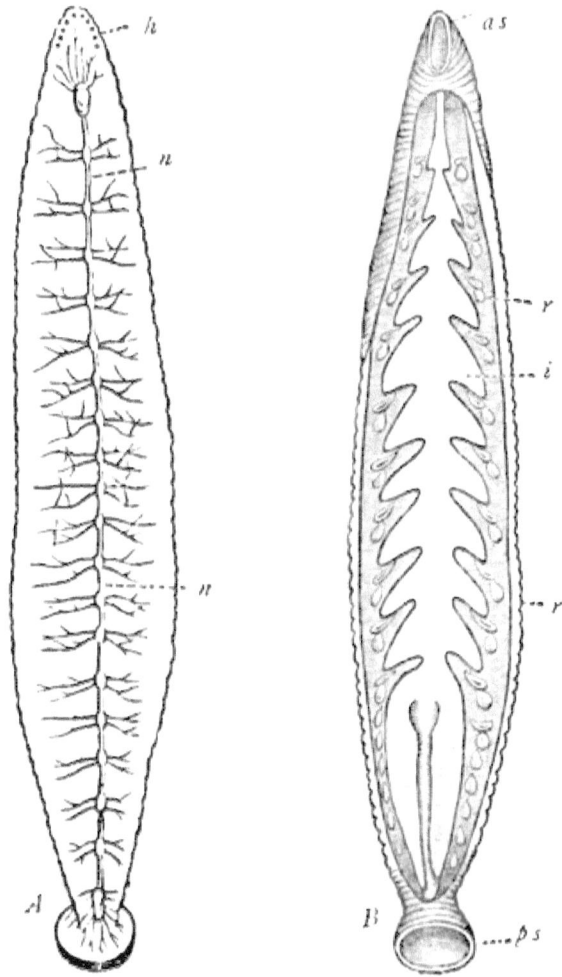

Fig. 11.—A, Diagram of the Leech, showing the nervous system, and the ten eyes placed on the top of the head. B, The leech dissected to show the alimentary canal (*i*), and the so-called "respiratory sacs" (*r r*); *as* Anterior sucker; *ps* Hinder sucker; *n n* Nervous system; *h* Head, carrying the eye-spots.

The mouth opens at the bottom of the front sucker, and is provided with three crescentic jaws (fig. 10, *b* and *c*), the edges of which are serrated with numerous minute teeth. These three jaws are so disposed as to meet in a single point, in a triradiate manner, and they are the organs by which the Leech cuts through the skin, in order to get at the blood-vessels beneath. Hence the bite of a leech consists of three little cuts radiating from a point.

The mouth opens by a short gullet into an alimentary canal, the first portion of which, usually termed the "stomach," occupies nearly the whole of the body-cavity, and is furnished with eleven membranous pouches on each side. The pouched "stomach" has the function of separating the watery part of the blood which the animal takes as food; and it opens into an intestine which terminates at a distinct vent placed on the back a little in front of the hinder sucker.

There is no distinct heart, and the place of the blood-vessels is taken by a peculiar system of tubes, containing a fluid which appears to play the same part in the economy of the animal as does the true blood of higher organisms.

It is doubtful, also, if there are any definite breathing-organs; but the animal is furnished with a series of little pouches (fig. 11, B, *r r*) which are placed on each side of the body, and have commonly been called "respiratory sacs." There are seventeen of these little organs on each side of the body, and they open on the lower surface by a series of minute pores.

The nervous system has the form of a chain of little nervous masses (fig. 11, A, *n n*) placed along the lower surface of the body, and united lengthways by cords. The first of these masses, representing the brain, is placed above the gullet, which is thus embraced by the cords which unite this with the next nervous mass in the series. The top of the head (fig. 11, A, *h*) also carries ten eyes, disposed in the shape of a horse-shoe.

The Leech attains a length of from two to three inches; its back is olive-green with rusty-red longitudinal stripes, and its lower surface is greenish-yellow spotted with black. In the nearly-allied *Sanguisuga medicinalis*, the lower surface of the body is olive-green, and is not spotted.

The Medicinal Leech is exclusively an inhabitant of fresh water, and is mostly imported into Britain from Hungary, Bohemia, Russia, and France.

RECAPITULATION OF ESSENTIAL CHARACTERS.—The body is elongated, and is ringed with transverse folds. When lateral appendages are present, these are never distinctly jointed or articulated to the body. The nervous system has the form of a chain of nervous masses placed along the lower surface of the body, and united by longitudinal cords. There is no distinct heart, but the circulatory system of the higher animals is represented, so far as its function is concerned, by a peculiar system of contractile tubes containing a coloured fluid. These characters distinguish the class of the *Annelida* as a whole.

CHAPTER IX.

CLASS CRUSTACEA.

THE class *Crustacea* comprises all the animals which are commonly known as Crabs, Lobsters, Shrimps, Prawns, Wood-lice, Water-fleas, and the like, all of which have the body enclosed in a more or less resistant shell. Hence they are sometimes erroneously spoken of as "Shell-fish;" and hence also their scientific name (Latin, *crusta*, a crust or shell). An admirable example of this class may be found in the common Lobster (*Homarus vulgaris*) of British seas.

The Lobster is almost completely enclosed in a strong and hard shell, which is really formed out of the skin, by the deposition in it of lime and horny matter. A little investigation also readily enables us to perceive that the animal (fig. 12) is really composed of a number of rings or distinct *segments*, placed one behind the other. Theoretically each segment is distinct, and each is constructed upon a similar plan or model. In theory, namely, each

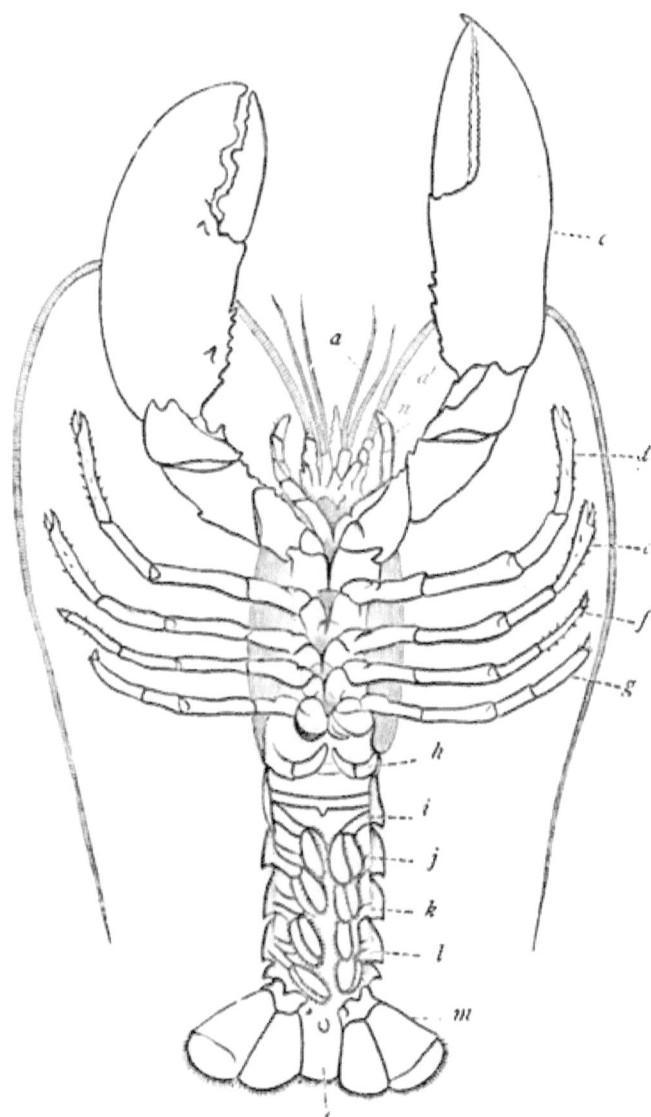

Fig. 12 — The common Lobster (*Homarus vulgaris*), viewed from below. *a* The lesser antennæ; *a'* The greater antennæ; *n* The last pair of foot-jaws; *c* The great claws, or first pair of legs; *d, e, f, g* The last four pairs of walking legs; *h, i, j, k, l, m* The six pairs of abdominal appendages, the last five being "swimmerets," and the last of all being greatly expanded; *t* The last segment of the body, without appendages.

segment consists of an upper and a lower arch of shell, joined to one another at the sides, and sending downwards a plate from the point where they join (fig. 13, 2). In theory, also, each segment carries a pair of "appendages," which ought to be composed of two branches springing from a common base. In some cases, as in the tail ("abdomen") of the Lobster, the actually existing segment really does conform to this theoretical type; but in other cases the segment may be very variously modified, and the appendages of the segment, in particular, assume the most different forms in different parts of the body. In practice, also, some of the segments are so amalgamated and consolidated with one another as to render their recognition a matter of great difficulty.

It will, however, greatly help us in studying the Lobster to remember that the body of the animal is really composed of a number of segments (twenty-one), each of which is constructed upon the theoretical plan or type just mentioned, and each of which may carry a single pair of appendages of some kind or other.

If we look at the Lobster from above (fig. 13, 1), we see at once that the body is very plainly divided into two parts, which would familiarly be called the "head" and "tail." The "head," as we should call it, is in truth composed of fourteen rings all amalgamated together, and covered above by a great shield or buckler (*ca*). The first seven of these segments belong to what is actually the head, whilst the hinder segments belong to what is properly called the trunk or "thorax" (Greek, *thorax*, a breastplate); and the head-shield shows this division by a transverse groove on its upper surface. On the other hand, the so-called "tail" is composed of a number (seven) of quite distinct rings or segments, which are movably jointed together, and which collectively constitute what is termed the "abdomen" (Latin, *abdo*, I conceal).

The Lobster is therefore really composed of twenty-one rings or segments, seven of which are free and movable, and constitute the abdomen, whilst the remaining fourteen are amalgamated together, but really belong, seven to a thorax and seven to a true head.

30   OUTLINES OF NATURAL HISTORY.

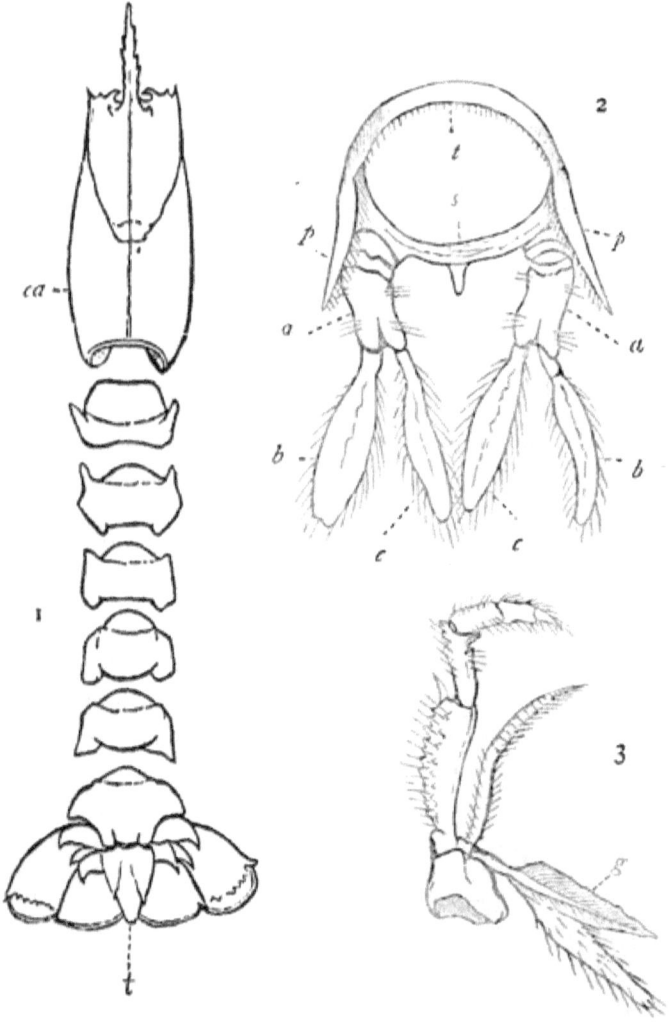

Fig. 13.—1. Lobster with all the appendages except the last pair of swimmerets removed, and the segments of the abdomen slightly separated from one another. *ca* The great shield covering the head; *t* The last segment of the abdomen left in connection with the last segment but one. 2. One of the segments of the abdomen, showing the typical structure of a segment and of the appendages; *t* Upper arch of the segment; *s* Lower arch; *p* Plate prolonged downwards from the line where the upper and lower arches unite; *a* Base of the appendage; *b* and *c* Outer and inner branches of the appendage. 3. One of the last pair of foot-jaws, carrying a gill (*g*).

Let us now briefly look at the various appendages carried on these segments, beginning at the head.

The *first* segment carries a pair of eyes, which are of large size and globular shape, and are supported upon movable stalks. The Lobster can thus roll the eyes about in different directions; and the eyes themselves are what is called "compound," each being composed of numerous simple eyes amalgamated together. The eyes also are protected in part by a great jagged spine or beak developed from the front of the shield which covers the head.

The *second* ring carries a pair of feelers, which are double, and composed of numerous joints (fig. 12, *a*). From their small size, these are known as the "lesser antennæ" (Latin, *antenna*, the yard-arm of a ship).

The *third* ring carries another pair of feelers (*a'*), which are known as the "great antennæ," from their large size. They are composed of numerous joints, like the preceding, but each consists of no more than a single branch.

The *fourth, fifth, sixth,* and *seventh* segments carry each a pair of jaws, differing somewhat in each segment, and the last pair so closely approaching the true legs or "feet" in structure as to receive the name of "foot-jaws." All these jaws move from side to side, and are really to be regarded as modified limbs.

The *eighth* and *ninth* rings (being the first two rings of the thorax) carry, each, another pair of limb-like jaws or "foot-jaws," the last pair of these (fig. 12, *n*) being of large size and quite like legs.

The *tenth, eleventh, twelfth, thirteenth,* and *fourteenth* segments carry five pairs of legs, which the animal uses partly in walking and partly for grasping. The first pair of these legs is greatly developed (fig. 12, *c*), and constitutes a pair of great pincers or "nipping-claws." One claw is blunt, and is used chiefly for holding on to foreign objects, and the other claw is sharply serrated, and is used by the animal for biting or cutting up its food. The second and third pairs of legs (*d* and *e*) also terminate in nipping-claws, but these are quite of small size. The fourth and fifth pairs of legs are provided with simply pointed extremities.

The *fifteenth* segment (being the first of the abdomen) carries (in the males) a pair of singular grooved processes (*h*), and the *sixteenth, seventeenth, eighteenth, nineteenth,* and *twentieth* segments carry each a pair of appendages which are known as "swimmerets" (*i, j, k, l, m*), and each of which consists of an undivided base, terminated by two flattened paddles or oars (fig. 13, 2). The last of these pairs of swimmerets (*m*) has the terminal paddles greatly expanded.

The *twenty-first* segment (*t*) carries no appendages, and is simply placed between the expanded swimmerets of the twentieth ring, thus constituting a most powerful tail-fin, by the strokes of which the animal can propel itself through the water, tail foremost, for an astonishing distance, and with great rapidity.

The following table shows the segments of which the Lobster is composed, with their proper appendages:—

HEAD,
- 1st, Eyes.
- 2d, Lesser antennæ.
- 3d, Greater antennæ.
- 4th, Pair of biting-jaws.
- 5th, First pair of chewing-jaws.
- 6th, Second pair of chewing-jaws.
- 7th, First pair of foot-jaws.

THORAX,
- 8th, Second pair of foot-jaws.
- 9th, Third pair of foot-jaws.
- 10th, First pair of legs (great claws).
- 11th, Second pair of legs (small claws).
- 12th, Third pair of legs (small claws).
- 13th, Fourth pair of legs (pointed).
- 14th, Fifth pair of legs (pointed).

ABDOMEN,
- 15th, Grooved appendages.
- 16th, Small swimmerets.
- 17th,   ,,   ,,
- 18th,   ,,   ,,
- 19th,   ,,   ,,
- 20th, Large swimmerets.
- 21st, No appendages.

We may add to this description of the Lobster, that not only is the body composed of a succession of segments, some of which are movably jointed to one another, but the limbs are also composed of distinct pieces or joints, and are movably jointed or "articulated" to the body. Hence the Lobster is emphatically what is called an "Articulate" animal (Latin, *articulatus*, jointed).

It remains very briefly to consider the more important points in the internal anatomy of the Lobster. The mouth is placed on the under surface of the head, and in addition to the numerous jaws already alluded to, it is further provided with an upper and lower lip, both of a shelly nature. The mouth leads by a gullet to a globular stomach, from which an intestine proceeds, to terminate by a distinct vent at the base of the last segment of the body; and there is a well-developed liver.

The heart (fig. 14, h) is placed upon the back, and drives the pure blood, which has passed through the gills, to all parts of the body. The breathing-organs ($b\ b$) are adapted for breathing air dissolved in water, and are therefore genuine "gills." They are in the form of pyramidal bodies, which are attached to the bases of the

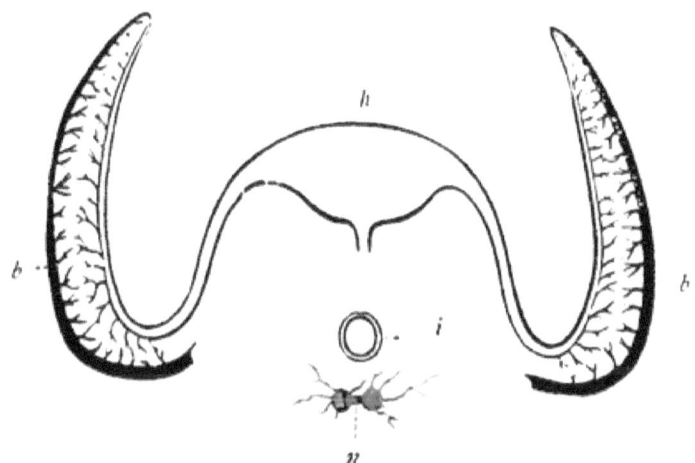

Fig. 14.—Diagram to show the position of the internal organs of the Lobster as they would be seen if the animal were cut across behind the head. *h* Heart; *b b* Gills (the vessels containing pure or arterial blood are left light, those containing venous or impure blood are dark); *i* Intestine; *n* Nervous system.

legs, and are concealed from view beneath the sides of the great shield which covers the head and thorax. Owing to their being attached to the legs, the animal's respiratory process depends very much upon its moving about, since the movements of the legs contribute considerably to the bringing in of fresh water to the chambers in which the gills are contained.

The nervous system, lastly, has the form of a series of nervous masses placed along the lower surface of the body, and united with one another by longitudinal cords. The first pair of these masses is placed above the gullet, and the cords which unite them with the next pair pass on each side of the gullet, so that the gullet is surrounded by a "nerve-collar."

The Lobsters are exclusively found in the sea; and though they can live a considerable time out of the water, they are essentially aquatic animals. They are exceedingly voracious, and are usually captured by means of "lobster-pots," or baskets baited with some kind of carrion or garbage. When injured, or even if greatly alarmed, they throw off one or both of the great claws; but these appendages soon grow again, though not so large as before. They also cast their shells periodically, since the resisting nature of this covering does not allow of their growth. When fresh they are very brightly coloured; but they turn to a uniform and brilliant red when boiled. They are most ordinarily about a pound in weight, but they sometimes grow to three or four pounds.

RECAPITULATION OF ESSENTIAL CHARACTERS.—The skin is hardened with lime and horny matter, so as to form a resisting shell or "crust," within which the internal organs are protected. The body consists of a succession of distinct rings or segments placed one behind the other; and each segment may carry a single pair of jointed appendages. The animal breathes air dissolved in water, and usually has breathing-organs in the form of gills. The nervous system consists of a chain of nervous masses placed along the lower surface of the body. Less essential, though highly distinctive, are the characters that the true legs are from five to seven pairs in number, that the segments of the abdomen carry appendages, that there are two pairs of antennæ, and that there is a distinct heart placed upon the back. By these characters the class of the *Crustacea* is distinguished as a whole.

CHAPTER X.

CLASS ARACHNIDA.

THE class *Arachnida* (Greek, *arachne*, a spider) comprises the Scorpions, Spiders, Mites, and Ticks. As an excellent example of this class we may select the common House Spider (*Tegenaria civilis*) of Britain. The body of this familiar animal (fig. 15, A) in reality resembles that of the Lobster, in being composed of a series of rings or segments, placed one behind the other; but these segments are not conspicuous, and the body only shows a well-marked division into two distinct portions—a front portion carrying the legs, and a hinder portion carrying no appendages.

The skin over the whole body is more or less hardened with horny matter; but more so in some parts than in others, and it nowhere forms a shell like that of the Lobster.

The front portion or half of the body (fig. 15, A, *c*) is in reality composed of the head and trunk ("thorax"), so consolidated together that no sign of a boundary between them can be made out, and that no distinct segments can be detected. On the sides of this region of the body we observe four pairs of long, jointed legs, and in front of these a pair of what look like small legs. These latter organs (*p*), however, are not really legs, but are a sort of feelers which are attached to the jaws. The spider, therefore, has truly four pairs of legs, and it should thus never be confounded with the genuine Insects, all of which possess no more than three pairs of legs. All the legs are long and slender, covered with numerous short hairs and a few longer spines, terminated by three claws each, and composed of seven distinct joints or pieces. They are fixed at their bases to a strong horny plate which covers the chest in front.

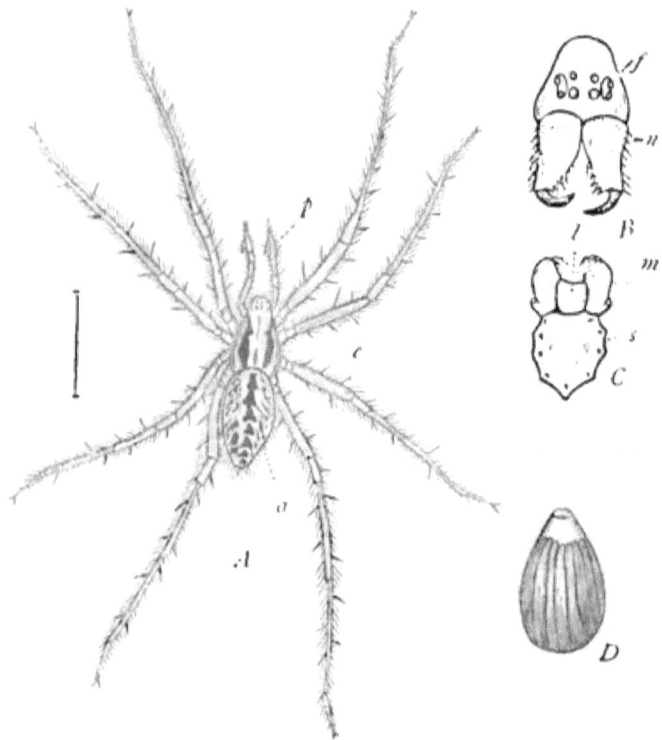

Fig. 15.—A, The male of the common House Spider (*Tegenaria civilis*), considerably magnified; *c* Front portion of the body, consisting of the amalgamated head and thorax; *p* Feelers attached to the jaws; *a* Abdomen. B, Front portion of the head of the same, showing the eight eyes (*f*), and the poison-jaws (*n*). C, Under side of the head and trunk, showing the true jaws (*m*), the lower lip (*l*), and the horny plate to which the legs are attached. D, Diagram of one of the air-chambers or breathing-organs. (Figs. A, B, and C are after Blackwall.)

Upon the front of the head are situated eight simple eyes (fig. 15, B, *f*), arranged in two lines, the pairs which occupy the ends of the line being placed on small tubercles. On the under surface of the head is situated the opening of the mouth, and in front of this are two organs, which are often spoken of as jaws, but which really correspond with the feelers ("antennæ") of insects. These singular organs (fig. 15, B, *n*) are in the form of two powerful jaw-like structures which terminate in strong curved hooks, and are movably jointed to the head. The hooks are perforated by a minute aperture at the point, communicating with a poison-gland; so that when one of these fangs is struck into the body of an insect, a drop of a poisonous fluid is forced out into the wound. In this way the spider kills the small animals upon which it feeds.

The mouth is closed behind by a plate representing the lower lip (fig. 15, C, *l*), and has at its sides two genuine jaws (*m*), which carry the jointed feelers already alluded to.

The front portion of the body of the Spider, as before remarked, is truly the amalgamated head and thorax. Behind this, and united with it by a narrow stalk, is the egg-shaped, hairy mass, which constitutes the hinder half of the body (fig. 15, A, *a*). This, though not exhibiting any distinct segments, is really the "abdomen," and corresponds with the "tail" of the Lobster. It does not support any legs or appendages, but at its hinder end are situated three pairs of minute conical eminences, which spin the fibres which compose the web, and which are termed the "spinnerets." The substance which composes the fibres out of which the web is constructed, is secreted within the body in a fluid form by certain special glands. The fluid silk is then cast into its proper thread-like form by being passed through the spinnerets, which are perforated by numerous very minute pores or holes. In this way the silken thread of the web is really composed of numerous very delicate filaments woven together.

Upon the under surface of the abdomen, far forwards, are situated two small openings which communicate with the breathing-organs. These latter have the form of little sacs or air-chambers, the lining membrane of which is thrown into numerous folds like the leaves of a book (fig. 15, D), and is richly supplied with blood. The air is admitted directly to these sacs, and the blood is thus purified; so that the Spider is an air-breathing animal.

As regards the internal anatomy of the Spider, little need be said. The digestive system presents no remarkable peculiarity, except that the throat is extremely narrow; a wide throat not being necessary for an animal which lives merely upon the juices of its prey. The intestine terminates in a distinct vent, and there is a well-developed liver. The heart has the form of a long tube placed along the back; and the nervous system has, in the young, the form of a chain of nervous masses placed along the lower surface of the body. In the adult, however, these masses are aggregated and amalgamated with one another.

Tegenaria civilis attains a length of nearly half an inch, and is the common House Spider of Britain. It spins a horizontal web, appended to which is a short open tube, into which the animal retreats when threatened with danger, and from which it watches for insects which may fall into its snare. It is mostly of a reddish-brown colour, with black markings. It lays its eggs in little packets of fifty or sixty each, all kept together by delicate silken fibres. Several of these packets are constructed and are attached to walls or other objects in the neighbourhood of the web. The young Spider is like its parent, but is much smaller; and it changes its skin no less than nine times before it assumes the characters of the adult. Its entire span of life appears to extend over four years.

RECAPITULATION OF ESSENTIAL CHARACTERS. — The body is composed of a series of rings placed one behind the other; and those rings which belong to the head and trunk (thorax) are amalgamated together. There are four pairs of legs. The abdomen never carries legs. The animal breathes air directly. The heart (if present) is situated on the back, and the nervous system has, at any rate to begin with, the form of a chain of little nervous masses placed along the lower surface of the body. The feelers ("antennæ") which are so characteristic of Insects, when present at all, are converted into offensive weapons (poison-fangs or pincers). These characters distinguish the *Arachnida* as a whole.

CHAPTER XI.

CLASS MYRIAPODA.

THE Myriapods are commonly known as Centipedes and Millipedes, and all these names refer to the great number of the feet or legs, as compared with Insects or Spiders, (Greek, *murios*, ten thousand; *podes*, feet). As a good representative of this class may be taken *Lithobius forficatus*, one of the commonest of British Centipedes.

The body in *Lithobius* (fig. 16, A) is elongated and flattened, and exhibits very distinctly a division into two regions—namely, a head (*h*), and a lengthy body composed of a series of separate rings or segments, each of which carries a single pair of appendages. The entire integument, both over the head and over the body, is hardened, so as to form a strong case, within which the internal organs are contained; and the appendages to the segments are very distinctly jointed.

The head is somewhat heart-shaped, and though it really consists of several pieces or segments, these are so consolidated that it appears to form a single piece. The head carries a single pair of long, jointed feelers, known as the "antennæ" (Latin, *antenna*, the yard-arm of a ship), which consist of very numerous short joints. The head also carries upon its two sides a collection of minute simple eyes (fig. 16, C, *e*), from twenty-two to twenty-four on each side. On its under surface the head bears the opening of the mouth, with a well-marked lower lip behind it—this latter organ being really double (fig. 16, B, *l*).

Immediately behind the head, and looking as if it belonged to it, is a narrow ring which carries a pair of powerful jaws (fig. 16, A, *f*, and B, *f*). These jaws are strongly hooked, and are perforated for the purpose of conveying a poison, with which the Centipedes kill their prey, or defend themselves against their enemies. Though officiating as jaws, these hooked fangs are really to be regarded as being modified legs or "feet," and they are, therefore, called "foot-jaws."

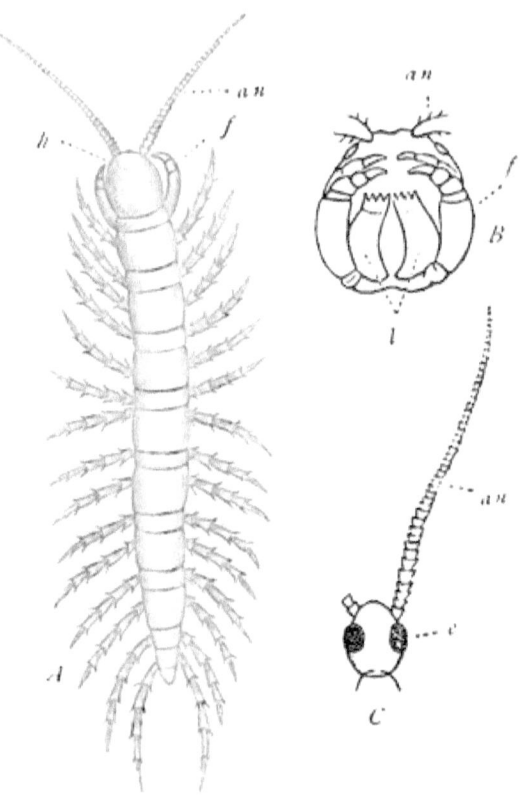

Fig. 16.—A, *Lithobius forficatus*, enlarged, and viewed from above; *an* Antennæ; *f* Foot-jaws; *h* Head. B, Head of *Lithobius Leachii*, viewed from below (after Newport); *an* Antennæ; *f* Hooked foot-jaws; *l* Lower lip, composed of two pieces. C, Head of *Lithobius forficatus*, viewed from above (after Gervais); *an* Antennæ; *e* Eye.

Behind the foot-jaws, the body exhibits fifteen distinct rings, alternately large and small (fig. 16, A), and each of these rings carries a pair of jointed legs. There are therefore fifteen pairs of legs, of which the hindmost are longer than the others, and are directed backwards in the line of the body, so as to form a kind of tail.

Little need be said as to the internal anatomy of *Lithobius*. The digestive tube is well developed, and is furnished with large salivary glands, and a rudimentary liver, together with certain tubes which represent the kidneys. The heart is in the form of a long, simple tube, placed along the middle of the back. The breathing-organs are in the form of delicate membranous tubes, the walls of which are strengthened by a spirally-coiled filament or fibre of horn. These tubes commence at the surface in little rounded apertures, one of which is placed on each side of each alternate segment, and they branch frequently as they proceed inwards amongst the various tissues of the body. The nervous system, lastly, has the form of a chain of pairs of little nervous masses, one pair being present in each ring or segment; these masses are united together so as to form a doubly-knotted cord, placed along the lower surface of the body.

Lithobius forficatus is a darkness-loving creature, and haunts obscure crevices of walls and cellars, or lives hidden under stones, or beneath the rotten bark of trees. It is highly carnivorous, and lives upon the bodies of small animals (such as earth-worms and caterpillars), which it kills by the bite of its empoisoned "foot-jaws." When captured, it will attempt to bite, but it is quite harmless, as the jaws are not sufficiently powerful to pierce the skin. Its general colour is brownish-red or ferruginous.

RECAPITULATION OF ESSENTIAL CHARACTERS.—Head distinct from the segments carrying the legs, and supporting a single pair of jointed feelers. Segments behind the head numerous and distinct, but not separated into regions. Legs numerous, usually from fifteen up to as many as one hundred and sixty pairs, sometimes eleven pairs, but never fewer than nine pairs. No wings. Breathing-organs in the form of branching tubes, adapted for breathing air directly. These characters distinguish the class of the *Myriapoda* as a whole.

CHAPTER XII.

CLASS INSECTA.

THE true Insects derive their scientific as well as their ordinary name from the distinctness with which the body is cut or divided into distinct portions or regions (Latin, *inseco*, I cut into). Of the many excellent representatives of this class which might be selected, none, perhaps, is better than one of the larger Dragon-flies, such as the great *Æshna grandis* of Britain (fig. 17, A).

If we look at a Dragon-fly, we observe very readily that the body is more or less clearly divided into three portions, a head in front, a chest (or "thorax") in the middle, and a tail (or "abdomen") behind. It will also be seen that the body is composed of a number of rings or "segments," which are placed one behind the other. These segments are very conspicuous in the tail or abdomen (fig. 17, A, a), are less conspicuous in the trunk or "thorax," and cannot be clearly discerned at all in the head.

The entire skin is hardened with horny matter, so that each ring or segment forms a more or less resistant tube, within which the internal organs are protected.

Commencing with the head (fig. 17, A, h), no distinct rings can, as already remarked, be clearly made out; but nevertheless the head really consists of a certain number of segments consolidated into a single mass. Upon its sides, the head has two conspicuous shining globes (fig. 17, D, e e), which are the eyes. Each eye is what is termed "compound," being really composed of an enormous number (several thousands) of minute eyes placed side by side, and doubtless conferring upon the creature a high power of vision. Besides these compound eyes, the head likewise carries three "simple" eyes, which are so minute as only to be visible with a magnifying glass. The head, further, carries upon its upper surface two jointed thread-like organs (fig. 17, D, *an*), which are called the "feelers" or "antennae" (Latin, *antenna*, the yard-arm of a ship),

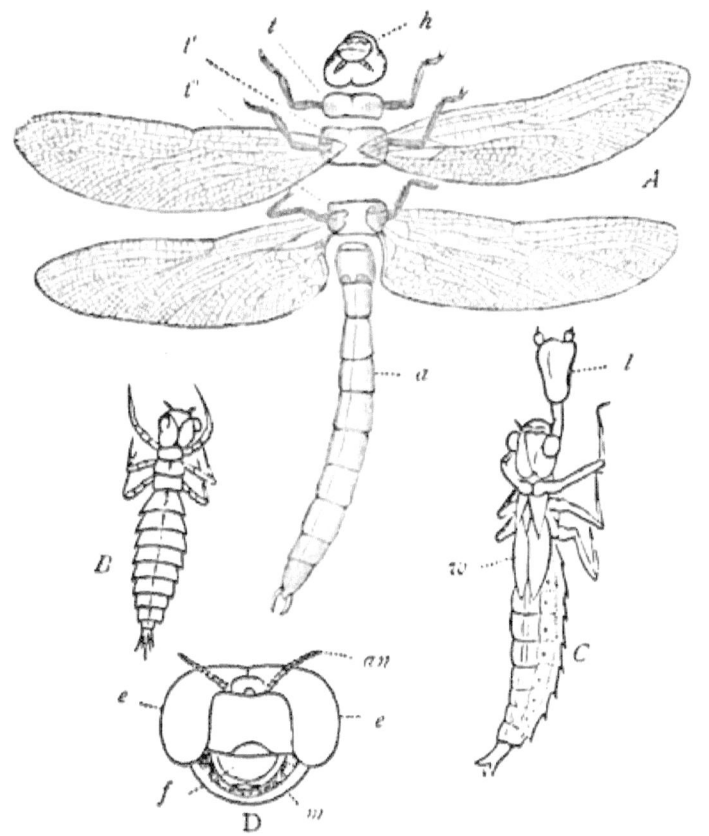

Fig. 17.—A, One of the Dragon-flies (*Æshna grandis*), slightly dissected; *h* Head, carrying the eyes, antennæ, and organs of the mouth; *t, t', t''* First, second, and third segments of the thorax slightly separated from one another, each carrying a pair of legs, and the two last carrying each a pair of wings; *a* Tail or abdomen. B, Young form, or "larva," of the same. C, Second stage, or "pupa." D, Head of a Dragon-fly (*Libellula depressa*), showing the feelers or antennæ (*an*), the eyes (*e e*), the hinder pair of jaws (*m*), and the upper lip (*f*).

and which no doubt are employed by the insect as organs of touch, and perhaps as organs of hearing as well. Upon its under surface, lastly, in front, the head carries the mouth, surrounded by the lips and jaws. These need not be particularly described, beyond saying that they consist of an upper and lower lip, and of two pairs of strong jaws. The jaws do not work up and down, as in man, but from side to side, and they are adapted for biting, thus enabling the insect to live upon other insects, which it captures and devours.

Behind the head come three rings, which are slightly separated from one another in the illustration (fig. 17, A, t, t', t''), but which in reality are consolidated with one another so completely, that they can only be made out by means of the appendages which they carry. They compose collectively a region of the body which is known as the chest or "thorax" (Greek, *thorax*, a breastplate). The first of these three rings (t) carries a pair of jointed legs; the second (t') carries another pair of similar legs, and a pair of wings; and the third (t'') carries a third pair of legs, and a second pair of wings. There are thus three pairs of legs and two pairs of wings.

The wings are really expansions of the skin, and are nearly of equal size. They are membranous, transparent, without hairs, and rendered gauzy by an extremely fine network of interlacing threads, which are known as "nervures." These threads are really hollow; and whilst they serve to support the fragile expansion of the wing, they also assist the insect in breathing, for they contain blood-vessels and prolongations of the breathing-tubes.

Behind the trunk or thorax comes the tail or "abdomen" (Latin, *abdo*, I conceal; so called because it conceals the internal organs). This region of the body is very distinctly composed of separate rings (nine in number), none of which carry legs, and all of which except the last are devoid of any appendages at all.

Turning now to the internal anatomy of the Dragon-fly, a few words may be said about its organs of digestion, its nervous system, and its breathing and circulatory apparatus. The mouth, armed with its powerful jaws, opens into a gullet (fig. 18, g), which conducts to a stomach (s). The stomach opens into an intestine (i), at the commencement of which are certain membranous tubes (f), which end in closed extremities, and which are believed to represent either the liver or the kidneys. The intestine opens into a large chamber (*c*), which opens upon the surface by a distinct vent (*v*).

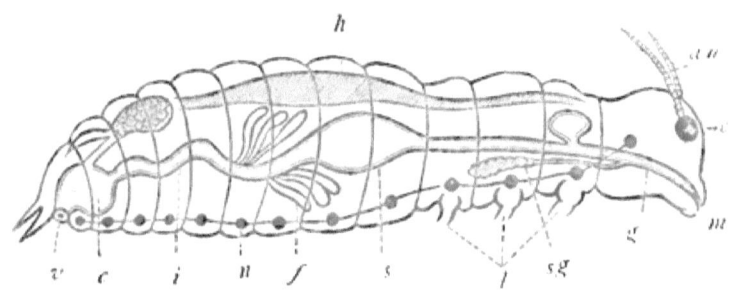

Fig. 18.—Diagram of the anatomy of an insect. *an* Antennae; *e* Eye; *m* Mouth; *g* Gullet; *sy* Salivary gland; *s* Stomach; *f* Tubes supposed to represent the liver; *i* Intestine; *c* Chamber into which the intestine opens; *v* Vent; *h* Heart; *n* Nervous system; *l* Bases of the legs.

The nervous system (fig. 18, *n*) consists of a chain of little nervous masses placed in pairs along the lower surface of the body, a pair of these masses being situated in each ring of the body. The first pair of nervous masses is placed above the gullet, and the second pair behind or below the gullet, and the cords which unite these two pairs pass on the sides of the gullet. It follows from this that the gullet is surrounded by a ring or "collar" of nervous matter.

The heart is in the form of a long tube (*h*), placed along the back, and furnished with flaps or valves, which only allow the blood to pass in one direction, namely, towards the head.

The breathing-organs are in the form of branched tubes, which commence on the surface of the body in little rounded openings, and then branch freely through the tissues, thus conducting the air to all parts of the body, and purifying the blood. The breathing-tubes are composed of a delicate membrane, the walls of which are supported by a horny fibre, which is coiled up in the interior in the form of a close spiral.

When young, the Dragon-fly is very different to the grown-up insect, and it passes through certain changes, before it assumes its final characters. The eggs are laid in the water, and the young insect, when hatched, presents the appearances seen in fig. 17, B. It exhibits a head, thorax, with three pairs of legs, and abdomen; but it shows no traces of wings. It swims about actively in the water, and devours smaller insects by means of its powerful jaws. It breathes by means of a tuft of valvular appendages placed at the end of the abdomen, which can be opened so as to allow the water to gain access to the intestine, the sides of which are furnished with folds containing within them breathing-tubes. When the blood has been thus purified, the water is thrown out from the intestine, and the jet thus produced drives the animal in the opposite direction. After a while the insect passes into a second stage (fig. 17, C), in which it resembles the preceding in most respects, but has rudimentary wings (*w*) placed upon the back of the thorax. It is still active and voracious. Again after a while, the animal drags itself out of the water, and climbs upon some plant. Its skin then dries, and splits along the back, and the perfect insect, with its fully developed wings (fig. 17, A) is set free, and flies away to lead an active existence in the air. These remarkable changes constitute what is known as the "metamorphosis" of the insect (Greek, *meta*, indicating change, and *morphe*, shape).

Æshna grandis is the largest of British Dragon-flies, attaining a length of about two and a half inches. Its general colour is yellowish-brown, with two yellow lines on each side of the thorax, and the abdomen variegated with green or yellow spots. At all periods of its life, and especially in its winged state, it is a most active and voracious insect, living upon other insects; in its final stage, in spite of its ferocity and destructiveness, it is one of the most beautiful and graceful of insects, from its large and brilliant eyes, its lustrous wings, and the ease and power with which it performs the most rapid evolutions in the air.

RECAPITULATION OF ESSENTIAL CHARACTERS. — The body is composed of a succession of rings, and is divisidle into three distinct regions—a head, thorax, and abdomen. The head carries a single pair of feelers (antennæ), the organs of the mouth, and the eyes. The thorax consists of three rings, and carries three pairs of legs. Generally, the last two segments of the thorax carry two pairs of wings. The adult insect breathes air directly, and the respiratory organs are in the form of branching breathing-tubes. The nervous system consists of a chain of little nervous masses placed along the lower surface of the body. The abdomen consists of distinct segments, which do not carry locomotive appendages. These characters distinguish the class of the *Insecta* as a whole.

CHAPTER XIII.

CLASS POLYZOA.

THIS class includes the curious Sea-mosses and Sea-mats, known technically as *Polyzoa* (Greek, *polus*, many; *zoön*, animal), because they consist of colonies or assemblages of little animals, associated so as to form compound growths—much in the way that a tree is composed of leaves and flowers supported upon a common trunk. As the type of this class we may take the broad-leaved Sea-mat or Hornwrack (*Flustra foliacea*), which is of common occurrence on the coasts of Britain.

This singular organism (fig. 19, *a*) is extraordinarily plant-like in form, and is generally regarded, when picked up on the shore, as being a pale-brown sea-weed. It forms a broad, and thin, leafy expansion, which is strongly rooted below by a common stem to a stone or some other foreign body, and which breaks up above into a number of flattened branches. Its consistence is horny, and its surface rough; and when it is examined with a magnifying glass, it is at once seen to be composed of an enormous number of little chambers or "cells," arranged in a single layer. These chambers (fig. 19, *b*) have horny

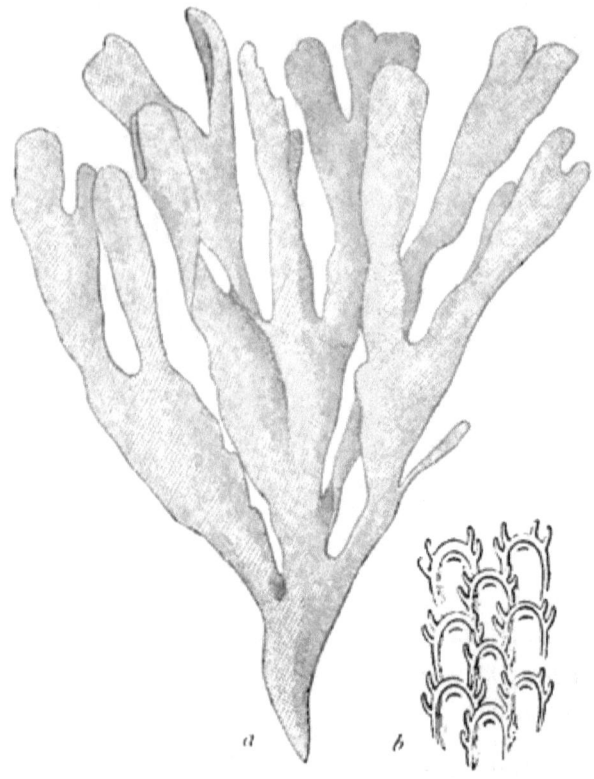

Fig. 19.—*Flustra foliacea*, one of the Sea-mats. *a* The plant-like colony, natural size; *b* A fragment of the colony magnified, showing the little chambers or cells, in which the separate animals forming the colony are contained.

walls, and are of an oval shape; each having a little transverse opening or mouth near its broadest end, and having its upper margin provided with four conical spines. Each cell contains a single animal, which leads an existence independent of the others; though the entire assemblage, made up of all the little animals contained in the innumerable cells, has also a life of its own, as a whole.

If we take a single cell of this compound growth, and examine the animal contained within it, we find that it has the following structure: The animal consists of a little membranous bag (fig. 20, *en*), which is closely applied to the horny wall of the cell (*ec*), and which is filled internally with a fluid, in which float the internal organs. From the mouth of the cell, the inner membranous sac can be partially thrust out; and here are situated the mouth and vent, close beside one another. The mouth (*o*) is surrounded by a circle of beautiful flexible processes (*t*), which are termed "tentacles," and the sides of which are fringed with innumerable vibrating hair-like filaments. These hair-like processes are in constant movement, vibrating to and fro, and by their means currents are set up in the surrounding water, and particles of food are thus conducted to the mouth. When quiescent, or irritated, the animal can draw in the front portion of its body, with the tentacles, into the shelter of the horny cell.

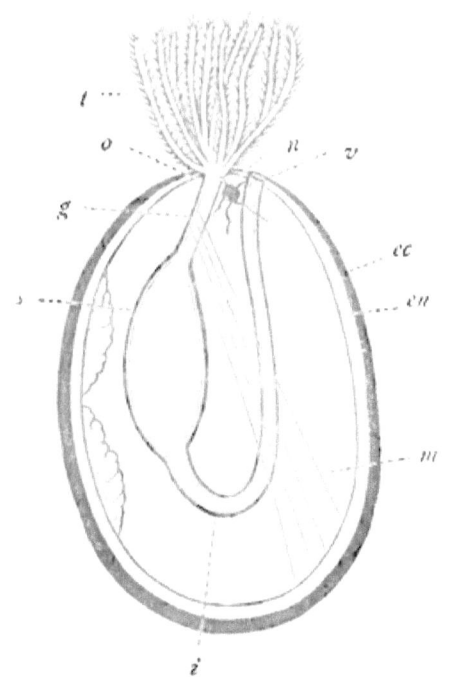

Fig. 20.—Diagram of the animal contained in one of the cells of *Flustra foliacea*. *ec* Outer horny wall of the cell; *en* Inner membranous wall enclosing the internal organs; *o* Mouth, surrounded by tentacles (*t*); *g* Gullet; *s* Stomach; *i* Intestine; *v* Vent; *n* Nervous system; *m* Muscle by which the animal can pull itself into its cell.

The mouth conducts by a gullet (*g*) into a stomach (*s*); and this in turn opens into an intestine (*i*), which finally terminates in a distinct vent (*v*), placed on one side of the mouth. To the gullet are attached certain muscular fibres (m), which are fixed below to the bottom of the cell ; and it is by the shortening of these that the animal pulls itself into the cell.

Between the mouth and vent is placed a little nervous mass, which constitutes the central portion of the nervous system (n). There is no heart, nor any blood-vessels ; nor are there any distinct breathing-organs. The fluid which fills the cell consists partly of water and partly of the products of digestion, and it is to be regarded as corresponding with the blood of higher animals. It is kept in movement by means of little vibrating hair-like processes which cover the membranous lining of the cell ; and it is exposed to the action of the oxygen contained in the water as it circulates through the crown of "tentacles," these latter organs being hollow.

The broad-leaved Sea-mat is of common occurrence in the seas of Britain, in a few fathoms of water. It attains a height of about four inches, and has a wood-brown colour. When fresh, it is stated to possess a peculiar odour (not always present), which has been variously compared to the scent of oranges, violets, or a combination of roses and geraniums, but which others consider strong and disagreeable (Johnston).

RECAPITULATION OF ESSENTIAL CHARACTERS.—Animal compound, consisting of numerous, nearly independent beings, each of which is enclosed in a separate chamber or cell. (This last character is not absolutely universal.) Each member of the compound growth has a mouth surrounded by tubular tentacles, a complete alimentary canal opening by a distinct vent, and a nervous system consisting of a little nerve-centre placed on one side of the mouth. There is no heart, nor are there definite breathing-organs. Almost universally the colony is attached to some foreign object. These characters distinguish the *Polyzoa* as a whole.

CHAPTER XIV.

CLASS TUNICATA.

THE animals known as *Tunicata* are all inhabitants of the sea, and derive their name from the fact that the body is enclosed in a sort of bag or "tunic." They are also often called *Ascidians* (Greek, *askos*, a wine-skin), from the fact that many of them have a close resemblance to a leathern wine-skin or bottle. "Rarely," remarks Prof. Edward Forbes, "is the dredge drawn up from any sea-bed at all prolific in submarine creatures without containing few or many irregularly-shaped leathery bodies, fixed to sea-weed, rock, or shell, by one extremity or by one side, free at the other, and presenting two more or less prominent orifices, from which, on the slightest pressure, the sea-water is ejected with great force. On the sea-shore, when the tide is out, we find similar bodies attached to the under surface of rough stones. They are variously, often splendidly, coloured; but otherwise are unattractive, or even repulsive, in aspect. These creatures are *Ascidiæ*, properly so called." As the type of this group we may take *Ascidia* (*Phallusia*) *mentula*, a not very rare inhabitant of British seas; but for various reasons it will be sufficient to indicate very briefly the chief points in its anatomy.

Ascidia mentula (fig. 21, A) presents itself in the form of an oval or oblong, leathery body, which is attached by the greater portion of one side to a stone or to a shell, and which has one extremity drawn out into two prominent necks, each perforated by a distinct opening. The animal thus has very much the appearance of a two-necked jar. If allowed to rest in a basin of sea-water, it can be seen that a current of water is drawn in by the longest or highest of the two necks of the sac or jar, and is expelled again from the lower neck. If the animal be rudely touched, it contracts itself forcibly, and squirts out a jet of water from both necks.

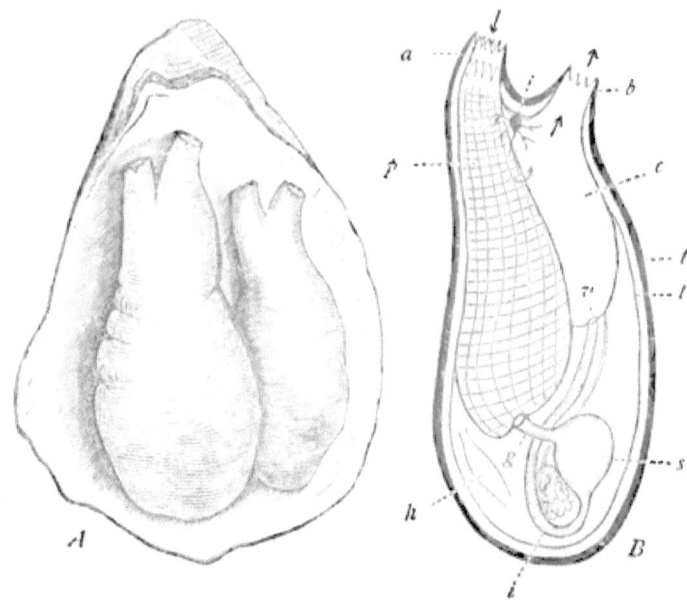

Fig. 21.—A, *Ascidia (Phallusia) mentula*. Two individuals attached to the inner surface of an Oyster-shell, natural size. B, Diagrammatic section of the same: *a* Opening by which the water enters; *b* Opening by which the water escapes; *p* Respiratory chamber; *g* Gullet; *s* Stomach; *i* Intestine; *v* Vent; *c* Chamber into which the water escapes after passing through the respiratory sac; *h* Heart; *t* Outer layer of the integument; *t'* Inner muscular layer or "tunic."

The outer covering of the body (fig. 21, B, *t*) is of a leathery nature, but is nevertheless partially translucent, especially in the neighbourhood of the necks. It is remarkable for containing a substance apparently identical with the woody fibre of vegetables. This outer covering is loosely lined by an inner layer or "tunic" (*t'*) which is of a highly muscular nature, and which confers upon the animal its power of squirting out water.

The highest of the two necks of the animal (fig. 21, B, *a*) is perforated by an opening surrounded by eight lobes, at the bases of which is a series of rudimentary eyes, in the form of yellow spots, each with a red point in its centre. This opening leads into a great chamber, which may be termed the "respiratory sac" (fig. 21, B, *p*), and which has a reticulated aspect, owing to its being perforated by numerous small apertures. The water which the animal takes in at *a* passes into this respiratory sac, and then escapes through the openings in its walls into a second chamber or sac, which opens on the surface at the other neck (*b*). In this way the creature both breathes and obtains its food. From the bottom of the perforated respiratory sac proceeds a gullet (*g*) which opens into a stomach (*s*), this in turn leading to an intestine (*i*), which terminates in a distinct vent (*v*) at the base of the second chamber.

There is a distinct heart (*h*) in the form of a tube open at both ends, and alternately propelling the blood in opposite directions. The nervous system is in the form of a little nervous mass (*n*) placed between the two necks.

Ascidia mentula attains a length of from one and a half to three inches, and may be obtained by dredging in from ten to thirty fathoms' depth. It can readily be kept alive with a little care; and it is sufficiently transparent to allow of the easy observation of the water-currents which constitute such a striking feature in its vital functions.

RECAPITULATION OF ESSENTIAL CHARACTERS.—Body furnished with two openings, conducting into two chambers which occupy the greater portion of the cavity of the animal. Internal organs enclosed in a double sac, the outermost layer of which is more or less leathery, whilst the inner is muscular. The nervous system is in the form of a single nervous mass. There is an alimentary canal, and a distinct heart. Respiration effected by currents of water which enter at one of the openings of the sac and are expelled from the other. These characters distinguish the *Tunicata* as a whole.

CHAPTER XV.

CLASS BRACHIOPODA.

This class comprises a number of shell-fish which agree with one another in having the body enclosed within a double ("bivalve") shell, and in having two long spiral processes, or "arms," attached to the sides of the mouth. From this last-mentioned peculiarity the name of the class is derived (Greek, *brachion*, arm; *podes*, feet). Owing to the great rarity of the few forms of Brachiopods which inhabit British seas, a foreign representative of the group has been selected—viz., *Terebratula flavescens* of the seas of Australia; and a brief description will suffice to indicate its leading peculiarities.

The body of *Terebratula flavescens* is soft, and the internal organs are enclosed in a modification of the integument, which is termed the "mantle." The front and back portions of the mantle produce a shell, which is composed of carbonate of lime, and conceals the animal within it. The shell (fig. 22, A) is double, or, in other words, consists of two distinct pieces, which are called "valves," so that the shell is said to be "bivalve" (Latin, *bis*, twice; *valvæ*, folding-doors). The entire shell has somewhat of the shape of an antique Roman lamp, each valve having a prominence or "beak" at one extremity, and one of them having this beak perforated by a round aperture (f), similar to the hole through which the wick passes into the lamp. (Hence the *Brachiopoda* are sometimes called "lamp-shells.")

The two valves of the shell are jointed together at the "beaks" by means of interlocking teeth and sockets, and they are of decidedly different sizes. The smaller valve lies over the back of the animal, and is not perforated by any aperture. Internally, however, it carries a singular shelly loop (fig. 22, B, l), which supports the

CLASS BRACHIOPODA. 55

bases of the spiral "arms," and which is sometimes called the "carriage-spring-apparatus." The larger valve lies

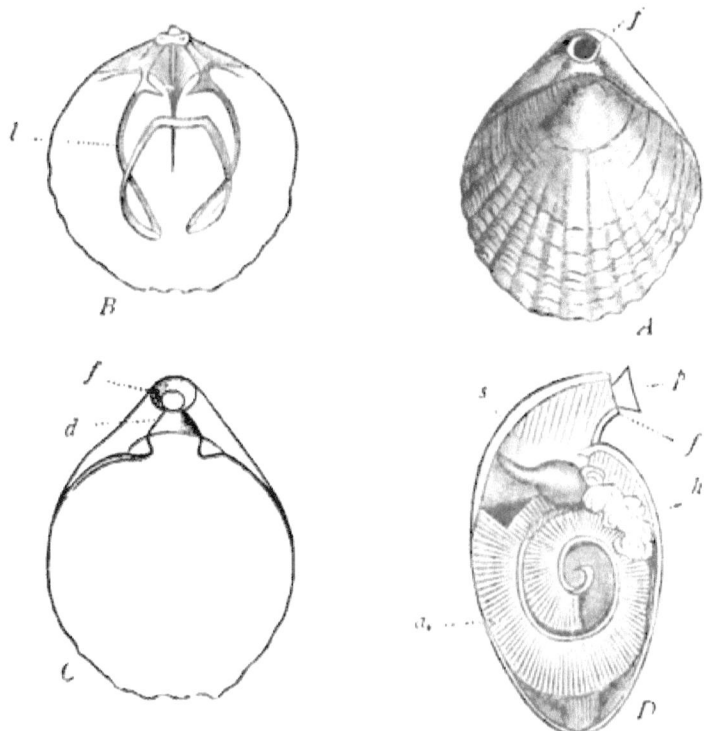

Fig. 22.—*Terebratula* (*Waldheimia*) *flavescens*. A, The shell viewed from behind, showing the smaller valve, and the perforated summit of the larger valve above it. B, Inner view of the smaller valve, showing the shelly loop (*l*) which supports the spiral arms. C, Inner view of the larger valve, showing the foramen or aperture (*f*) in the beak, through which the muscular stalk of attachment passes. D, Longitudinal and vertical section of the animal, showing the spiral arms (*a*), the stomach (*s*), and the liver (*h*). At *f* is the opening in the beak, with the stalk of attachment (*p*) passing through it. After Davidson and Owen. Some details have been omitted in figs. B, C, and D, for the sake of clearness.

upon the lower surface of the animal, and has its beak perforated by a large rounded aperture (fig. 22, C, *f*). Through this opening passes a muscular stalk, by which the shell is firmly fixed to some foreign body (fig. 22, D, *p*).

The animal contained within this shell exhibits many peculiarities of structure, but none is more striking than the fact that there springs from the mouth a pair of long, flexible, fleshy processes, which carry numerous filaments on one side, and are closely coiled up into a spiral (fig. 22, D, *a*). These singular processes are partially supported by the shelly loop (fig. 22, B, *l*) already spoken of as existing in the smaller valve of the shell. They are termed the "arms," and it is by their instrumentality that particles of food are brought to the mouth. The animal possesses a well-developed stomach (*s*), and an intestine, the latter terminating blindly. There is also a large liver (*h*). The nervous system consists of a central mass placed near the gullet.

Terebratula flavescens inhabits the seas of Australia, and lives a sedentary life, being attached to submarine objects by means of a muscular stalk which passes through the aperture in the beak of the larger valve of the shell.

RECAPITULATION OF ESSENTIAL CHARACTERS.—Animal included in a bivalve shell. The valves of the shell very markedly different in size (sometimes very slightly so), and placed over the back and front of the animal. Shell attached to some submarine object by a muscular stalk (sometimes by the shell itself). Mouth furnished with a pair of long, spirally-coiled, fringed processes or "arms." These characters distinguish the class of the *Brachiopoda* as a whole.

CHAPTER XVI.

CLASS LAMELLIBRANCHIATA.

THIS division includes the numerous animals commonly known as "Bivalve shell-fish," such as Cockles, Mussels, Oysters, Scallops, and the like. These animals derive their common name from the fact that they have the body protected by a shell, which is composed of two pieces or

CLASS LAMELLIBRANCHIATA.

"valves" (Latin, *valvæ*, folding-doors), and which is for this reason said to be "bivalve." They derive their scientific name, on the other hand, from the fact that they breathe by means of gills (Greek, *branchia*, a gill), and that these gills have a flattened, plate-like, or "lamellar" form (Latin, *lamella*, a flat plate).

As the type of this class we shall take the great Sand Gaper (*Mya arenaria*); not because it is especially common, but because it exhibits particularly well many of the characters of the Bivalves. The body of the Gaper is quite soft, and may be regarded as a kind of sac or bag, formed by the skin. This bag is termed the "mantle," because it conceals and protects the internal organs within it like a cloak. The bag formed by the mantle has two openings into it. One of these openings is placed at the hinder end of the body, and serves

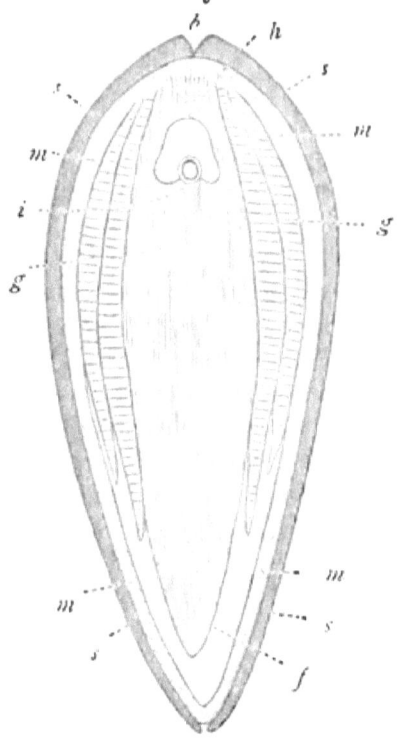

Fig. 23 — Diagrammatic vertical and transverse section of *Mya arenaria*. *b* Back, or "dorsal margin" of the shell; *s s* The two valves of the shell, right and left; *m m* The two halves, or "lobes," of the mantle, producing the shell; *g g* The gills, two pairs on each side; *h* The heart; *i* Intestine; *f* The foot.

as an aperture by which water is admitted into the interior of the body. At this opening the sac is drawn out into a long tube (fig. 24, *s*), which is really double, and which the animal can thrust out and draw in again. The other opening into the mantle-sac is for the purpose of allowing what is called the "foot" to be thrust out—this really being a muscular, tongue-shaped organ, by means of which the animal can shift its position (fig. 24, *f*).

The mantle, though forming in this way a closed bag, with no other openings into it except those just mentioned, really consists of two halves—a right and left half—and each half produces upon its outer surface one of the "valves" of the "shell" (see fig. 23, where *m m* are the two halves of the mantle, and *s s* the two valves of the shell). If, therefore, we take the animal in its living state, we do not see any portion of the body, except the long water-tubes already spoken of; but we see the double or "bivalve" shell. As the two halves of the mantle are right and left, and as each produces one valve of the shell, it follows that the shell consists of a "right valve" and "left valve." We may therefore compare the Gaper to a man enclosed within two great shields, one placed upon his right arm and one upon his left arm.

Leaving the shell, however, for the present, let us now examine the internal anatomy of the Gaper. In order to do this it is necessary to remove one of the valves of the shell, and we cannot do this without some violence to the animal within. The valve, namely, is attached to the mantle which produces it, and is also kept in firm connection with the opposite valve by means of two strong muscles, which are known as the "adductor muscles" (Latin, *adduco*, I lead towards or bring together). Hence in taking off one valve, we have to cut the mantle along the line where it is attached to the shell, and also to cut the two adductor muscles. When this is done, and the valve and mantle on one side are removed, we have the appearances presented in fig. 24. In this figure, the cut edge of the mantle is seen at *m*; the letter *a* represents the front adductor muscle, which has been cut through; and *a'* is the hinder adductor muscle, which has also been divided. We know that *a* is the *front* adductor muscle, because close beside it is situated the mouth (*o*); and the mouth, of course, is placed on the front of the body. The mouth is surrounded by four long membranous leaf-like processes or feelers (*p*). Immediately above the mouth, and occupying the greater portion of the centre of the figure, are two flattened membranous plates (*b*),

CLASS LAMELLIBRANCHIATA. 59

one being nearly hidden behind the other. These are the gills upon one side, and they constitute what in the case of the oyster is termed the "beard." Two similar gills are present on the other side of the body, but these are concealed from view. To the left of the gills we see the heart (h), the last portion of the intestine (v), and the hinder adductor (a'). To the right of the gills, below the mouth, we see the tongue-shaped muscular organ which is known as the "foot" (f). Lastly, above the gills, at the hinder end of the body (the end opposite to the mouth) we see the two muscular tubes (the so-called "siphons") by which water is admitted to the interior of the body, and again expelled.

We may examine the internal organs of *Mya* a little more minutely by the help of the following diagram (fig. 25). The *Mya* possesses no distinct head, but this end of the body

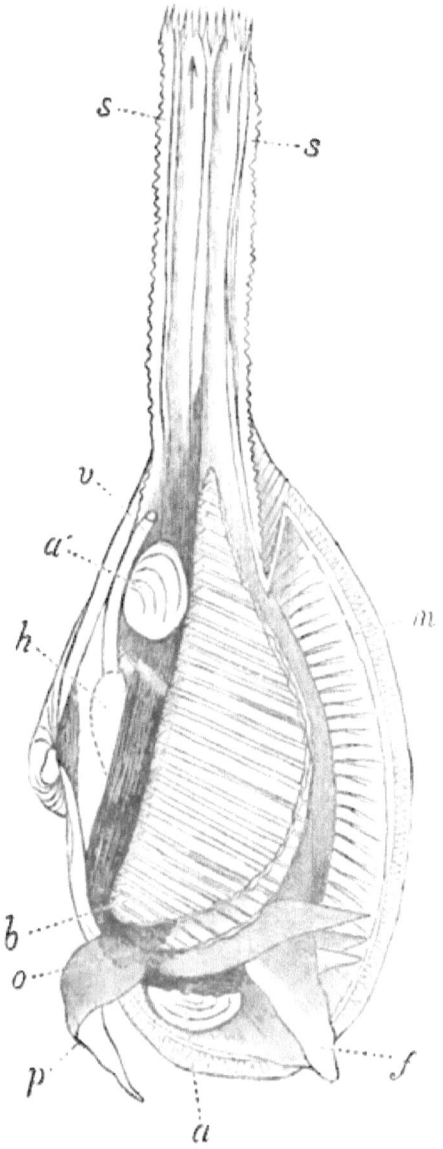

Fig. 24.—Anatomy of *Mya*, after one valve and one half of the mantle have been removed. m Cut edge of mantle; $s\ s$ The breathing-tubes (siphons) cut in half; a Front adductor; a' Hinder adductor; h Heart; o Mouth, surrounded by four membranous processes or feelers; f Foot; v Vent; b Gills.

can be recognised by the presence of the mouth (*o*) with its membranous feelers. The mouth is destitute of teeth,

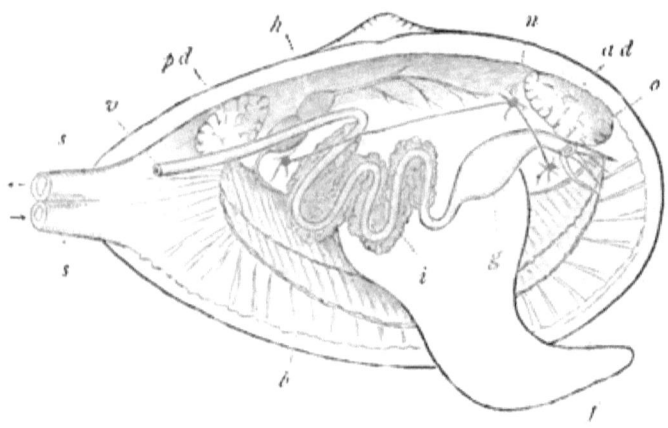

Fig. 25.—Diagram of the anatomy of *Mya*. *o* Mouth ; *g* Stomach ; *i* Intestine, surrounded by the liver ; *v* Vent ; *b* Gills ; *h* Heart ; *s s* Breathing tubes ("siphons") ; *f* Foot ; *n* Nervous system ; *ad* Front adductor muscle ; *pd* Hinder adductor muscle.

and leads through a gullet into a stomach (*g*). From the stomach proceeds a long winding intestine (*i*) which is surrounded by a well-developed liver, and which finally terminates in a distinct vent. The vent is placed at the hinder extremity of the body, and is so situated as regards the breathing-tubes (siphons) that all undigested particles of food are carried away by the outgoing current of water which has passed over the gills. The nervous system (*n*) consists of three little nervous masses, connected by cords. There is a distinct heart (*h*), which drives the pure blood, which has come from the gills, to all parts of the body. The gills are in the form of membranous plates (*b*), two on each side of the body, and having their surfaces covered with minute hair-like processes, which lash to and fro in constant vibrations, and sweep the water over the gills.

We are thus led to consider how the water reaches the gills, for the animal would die of suffocation unless it could constantly get a supply of fresh water. In order to understand this, it is necessary to know how the *Mya*

CLASS LAMELLIBRANCHIATA.

lives. The animal lives buried in the sand, and buried head downwards, so to speak—that is to say, the mouth is turned downwards, and the end where the "siphons" or breathing-tubes are, is turned upwards (as in fig. 24). The breathing-tubes are two long muscular canals, which are so united to one another as to look like one tube, though really quite distinct internally. They can be thrust out of the shell at will, and again partially withdrawn within the shell by means of proper muscles. When the animal wishes to breathe or obtain food, it thrusts out these breathing-tubes through the sand in which it is buried till they reach the water above. Then the water is drawn in through the mouth of one of the tubes in a constant current and is carried to the gills in the interior of the body (see figs. 24 and 25, where the direction of the water-currents is indicated by arrows). Having passed over the gills, and purified the blood in its passage, the

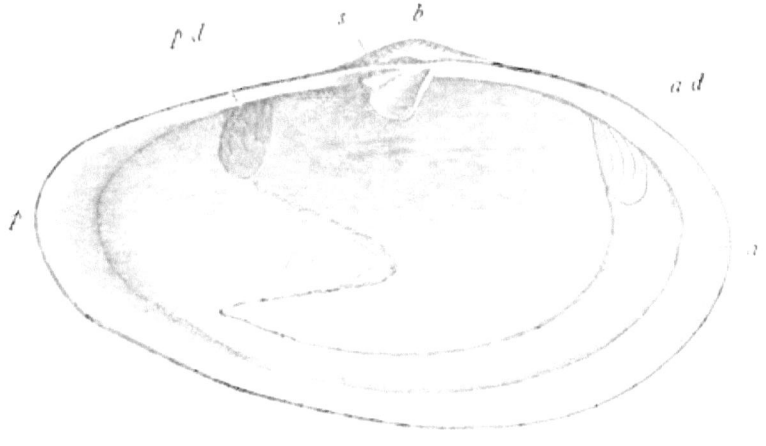

Fig. 26.—Interior of the left valve of *Mya arenaria*. *b* Beak; *a* Front end of the shell; *p* Hinder end of the shell; *ad* Scar or impression of the front adductor muscle; *pd* Impression of the hinder adductor; *ps* Line where the muscles which pull the siphon in are attached; *pl* Line where the mantle is attached; *s* Spoon-shaped process carrying the "cartilage."

water next reaches the mouth, and the animal extracts from it all the floating particles of food which it may contain. Finally, the water is conducted in a reverse direction along the intestine, taking with it all undigested food, till it reaches the second breathing-tube, from which it is expelled in an outgoing current. On the other hand, when the animal is disturbed, or when the sand in which it lives is left bare by the retreating tide, the breathing-tubes are partly drawn into the shell, and these currents cease.

We may now shortly examine the shell of *Mya*. As before said, the shell consists of two pieces or *valves*, which, as regards the animal within, are *right* and *left*.

The two valves are like each other in general form, but the left valve is slightly the smallest. The valves are hollow or concave, and they are applied to one another along their concave aspects. They do not, however, fit quite closely, but leave an opening for the breathing-tubes, so that the shell is said to "gape." Each valve is furnished along its "dorsal" margin with a prominent process or "beak" (fig. 26, *b*), and the valves are so applied to one another that the beaks are nearly in contact and are opposite one another. Between the beaks is a mass of horny fibres which are compressed when the valves are closed, so that when the animal relaxes the adductor muscles (by which the shell is closed), the valves are elastically forced apart. These horny fibres constitute what is called the "ligament" and "cartilage," and they are carried in part by a spoon-shaped process of shell developed below the beak of the left valve (fig. 26, *s*).

Externally, the shell exhibits numerous fine lines, which run concentrically round the beaks, and which mark the stages of the growth of the shell. The shell is also covered with a thin, brown, or reddish-brown membrane, which gives it its colour.

The form of the shell is an oval, broader at one end than the other. The beak is not placed quite in the middle of the shell, but somewhat to one side. This side of the shell is the broadest and shortest, and is the side on which the mouth is situated (the "anterior" or front side, fig. 26, *a*). The side of the shell from which the beak turns away is the longest and narrowest, and is the side at which the vent is situated (the posterior or hinder side, fig. 26, *p*).

Internally, the shell exhibits several points of importance. Placed beneath the "dorsal" margin of the shell (that is, the margin on which the beaks are situated), are two depressed and smooth impressions. One of these (fig. 26, *ad*) is placed near the mouth at the front end of the shell, and marks the point where the front adductor muscle was attached in the living animal. The other (fig. 26, *pd*) is placed towards the hinder end of the shell, and indicates where the hinder adductor was attached. Running from the one of these impressions to the other is a well-marked line (fig. 26, *pl*), which takes a course a little within the margin of the valve, and has a deep indentation (*ps*) opposite to the hinder end of the shell. This line marks the place where the mantle was attached to the shell, and the indentation or bay marks the point where the muscles which pull in the siphon were attached to the shell.

It follows from the preceding that if merely shown a single valve of the *Mya*, and knowing nothing of the animal, we should be able to state the following points: 1. That the broadest end of the shell was the one where the mouth was situated, because the beak turns to this end, and this half of the shell is the shortest. (This would not always be true of all Bivalves.) 2. That the animal possessed two adductor muscles for closing the shell. 3. That the animal had breathing-tubes or siphons for conducting the water to the gills, and that these tubes could be partially withdrawn within the shell.

Mya arenaria is found at various points along the British coast, imbedded in sand or mud, generally on long stretches of nearly level shore which are only uncovered at spring-tides. When the tide is out, the position of the shell is indicated by a rounded or oval hole, from which the animal squirts out water when the foot is put down near it. The shell is situated about five or six inches below the surface, with the siphons pointed upwards; and the animal has the power of shifting its position within its burrow by protruding its "foot"—this, as already said, being a tongue-shaped muscular organ (fig. 24, *f*), which can be thrust out between the valves of the shell. The length of the shell may be over three or four inches, with a breadth of about two inches and a half.

RECAPITULATION OF ESSENTIAL CHARACTERS.—The body is soft, and is enclosed in a skin or integument which is termed the "mantle." The mantle produces a "shell," which protects the soft body within, and which consists of two pieces or "valves," placed one upon the right side and one upon the left side of the body. There is no distinct head, and the mouth is destitute of teeth. The breathing-organs are in the form of lamellar or plate-like gills, disposed on the sides of the body. These characters distinguish the class of the *Lamellibranchiata* as a whole.

CHAPTER XVII.

CLASS GASTEROPODA.

IN this class are included all those animals which are commonly called "Univalve Shell-fish," such as the Whelks, Snails, Periwinkles, Limpets, &c. The name of "Univalves" is applied to them because most of them possess a shell which is composed of a single piece or "valve;" and they derive the name of *Gasteropoda* from the fact that the lower surface of the body is generally flattened out so as to form a broad expansion or disc, which is called the "foot," and upon which the animal creeps about (Greek, *gaster*, belly; *podes*, feet). As the type of this class we may select the common Whelk (*Buccinum undatum*) of British seas.

If we examine a Whelk whilst living and active, we

observe that it is a slug-shaped animal, which walks, or
rather creeps about, upon the lower surface of its body,

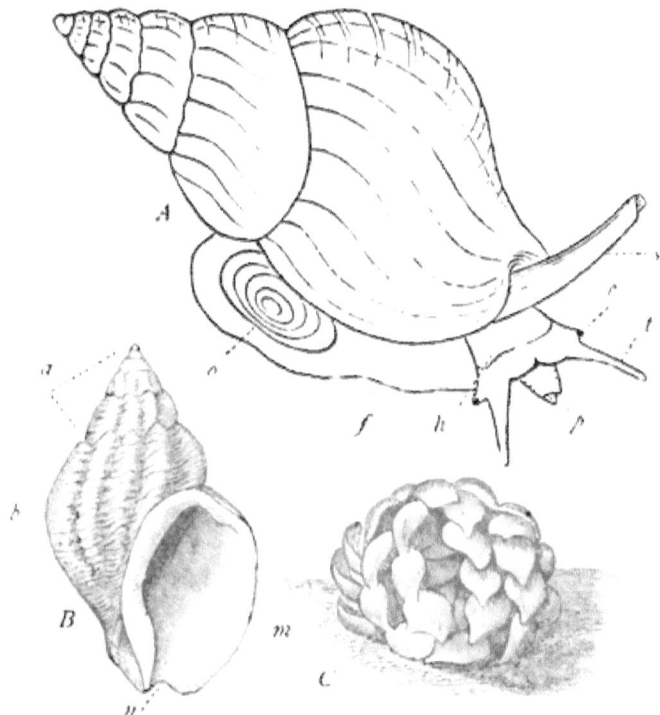

Fig. 27.—A, Sketch of a Whelk *Buccinum undatum*) in motion; *f* Foot;
h Head carrying the feelers (*t*) with the eyes (*e*) at their bases; *p* Proboscis; *s* Respiratory siphon, or tube by which water is admitted to the gills; *o* Operculum. B, Shell of the Whelk: *a* Spire; *b* Body-whorl; *n* Notch in the front margin of the mouth of the shell; *m* Outer lip of the mouth of the shell. This figure is half the natural size. C, A small cluster of the egg-capsules of the Whelk. (B and C are after Woodward.)

carrying over its back a shell composed of a single piece
(fig. 27, A). The surface upon which the animal creeps
forms a flattened and very muscular disc, which is termed
the "foot" (fig. 27, A, *f*). In front the animal exhibits a very distinct head (*h*), which carries a pair of
extensible processes (*t*), the purpose of which is to act as
"feelers," or organs of touch. At the bases of these
feelers are situated the two eyes, in the form of small

coloured spots. Immediately above the head is a folded tube, which the animal can thrust out to a considerable length. This tube (s) acts as a pipe or "siphon," by which fresh water is carried to the gills. At the hinder end of the foot we observe an oblong horny plate (o) with numerous concentric lines upon it. This plate is known as the "operculum" (Latin for a lid), and the function which it discharges is obvious. When the animal, namely, retires into its shell, the portion of the foot which carries this horny plate is the last to be drawn in; and as it fits accurately into the mouth of the shell, the Whelk is thus protected by the operculum against injury.

Over its back the Whelk carries a *shell*, which is composed of a single piece, and is therefore said to be "univalve." The shell has the general form of a cone, the broadest end of which is turned toward the head of the animal, whilst the pointed end is directed backwards. In reality, the shell (fig. 27, B) is composed of a conical tube, which is twisted in an oblique manner round a central pillar, and which therefore forms a "spiral." The first few turns of the shell are comparatively small, and they constitute what is termed the "spire" (fig. 27, B, a). The last turn of the shell is by far the largest (fig. 27, B, b), and as it contains the greater portion of the body of the animal, it is termed the "body-whorl." All the turns of the shell are in contact with one another, and the body-whorl opens in front by a large oval aperture, from which the animal can protrude itself, and which is known as the "mouth" of the shell. The inner margin of the mouth is formed by the pillar round which the whole shell is coiled, and to this pillar the animal is firmly attached by means of a special muscle. Lastly, the mouth of the shell exhibits *in front* a well-marked notch (fig. 27, B, n), which is for the passage of the breathing-tube or "siphon" already spoken of.

Externally, the shell exhibits numerous lines or striæ running parallel with one another and with the spiral turns of the shell. There is also a series of undulations or folds which run in the direction of the length of the shell, or, in other words, from the mouth towards the apex of the spire. The outer surface of the shell is of a brownish-white or yellowish-brown colour, whilst the mouth is white or flesh-coloured.

Turning now to the internal anatomy of the Whelk, the mouth (of the animal, not of the shell) is found to contain a singular organ which is known as the "tongue," and a portion of which is represented in a highly magnified form in fig. 28. The tongue consists of a long strap, which carries three rows of minute serrated teeth, composed of flint. By means of proper muscles this toothed strap can be made to move backwards and forwards over a kind of cushion upon which it rests. The animal can thus apply it like a saw to any foreign substance, and as the teeth are extremely hard, holes can be bored into other shells with great readiness. The mouth (fig. 29, *a*) is placed at the end of a proboscis, which can be thrust out to a considerable distance, and conducts by a gullet to a proper digestive cavity or stomach. From the stomach proceeds a long convoluted intestine, in great part surrounded by a voluminous liver. The intestine (*d d*) terminates in a distinct vent (*e*), which is placed upon the back.

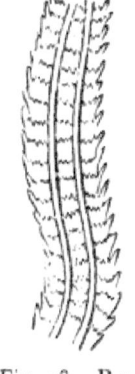

Fig. 28.—Portion of the tongue of the Whelk, highly magnified (after Woodward).

The nervous system is chiefly aggregated round the gullet (fig. 29, *f*). There is a distinct heart (*h*) consisting of two cavities or chambers. The breathing-organs are in the form of two plume-like gills (*g*), which are placed in a sort of chamber, formed by the folding of the mantle, on the back of the animal. The water necessary for respiration is admitted to this chamber by means of the folded tube or "siphon," which has been mentioned as being protruded through the notch in the front of the shell.

The common Whelk is very widely distributed throughout European seas, and is one of the most abundant of British Univalves. It is found usually in tolerably deep water, but extending from low water to depths of one hundred fathoms. It is very voracious, and exclusively

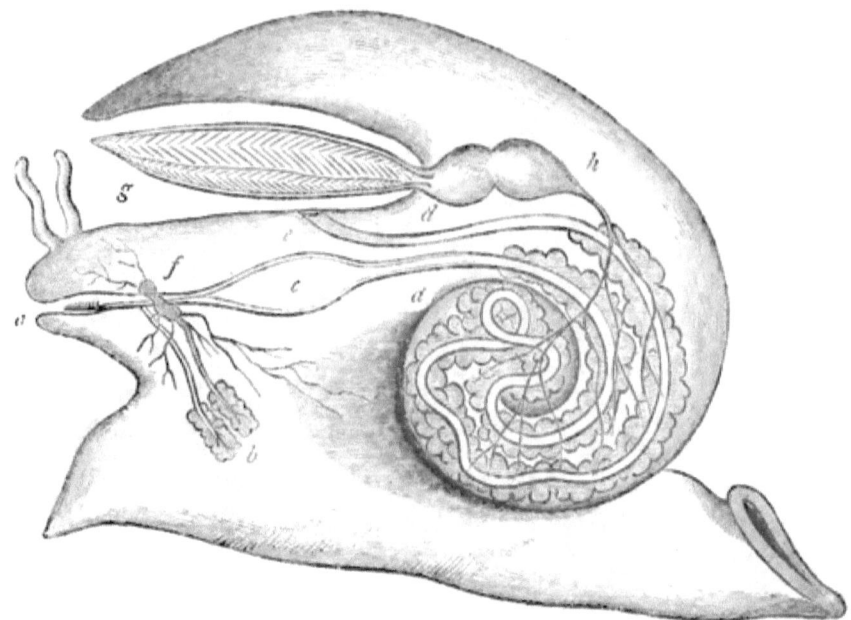

Fig. 29.—Diagrammatic section of a Whelk. *a* Mouth, with masticatory apparatus or tongue; *b* Salivary glands; *c* Stomach; *d d* Intestine, surrounded by the liver, and terminating in the vent (*e*); *g* Gill; *h* Heart; *f* Nervous system.

carnivorous, living upon other shell-fish or upon any dead animal bodies. It usually bores its way into other shells by means of the toothed tongue. It is not uncommonly used by fishermen as bait, and it is also occasionally eaten. The female Whelk lays its eggs in clusters of flask-shaped, horny capsules, each capsule containing five or six eggs. These clusters (fig. 27, C) are attached to stones, shells, or other foreign bodies, and the young, after attaining a certain degree of development, escape from the capsules by means of rounded perforations in the sides of the latter.

RECAPITULATION OF ESSENTIAL CHARACTERS.—Of the

above characters the ones which the Whelk shares with the other Gasteropods, and which therefore characterise this *class* of shell-fish, are only the following: The head is distinctly marked out from the rest of the body; the mouth is furnished with a peculiar toothed apparatus or "tongue;" the "foot" is used for locomotive purposes, having the form (not always) of a broad, flattened, muscular disc; and the body is not enclosed in a bivalve shell. Non-essential, though very common, is the character that the soft body is protected by a shell which is "univalve" or consists of a single piece.

CLASS PTEROPODA.

This class of animals contains minute shell-fish, which are found swimming in the open ocean far from land. They derive their name of *Pteropoda* from the fact that the head is furnished with two wing-like fins, by means of which the animal swims (Greek, *pteron*, wing; *podes*, feet). We may select as the type of this class the well-known *Hyalea tridentata* of the Mediterranean and Atlantic, but we shall give merely a brief outline of the most important points in its organisation.

The animal of *Hyalea tridentata* (fig. 30) is enclosed in a small, yellowish-brown, semi-transparent shell, which may be regarded as composed of a back and front plate, united to one another more or less completely. The back plate is nearly flat, and is prolonged in front so as to form a sort of hood. The front plate is strongly rounded and globular. Behind, the shell is prolonged into three spines, which arise from the line where the two plates of the shell unite with one another. In front the two plates leave a small aperture, through which the animal can protrude its head at will; and at the

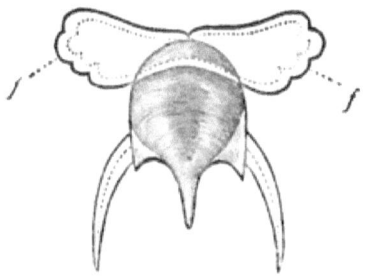

Fig. 30.—*Hyalea tridentata*, showing the shell and the lateral fins attached to the sides of the head (*f f*).

sides there are two slits, one on each side, through which pass long appendages of the integument or "mantle."

The animal is attached to the interior of its shell by a muscle which passes from the point of the shell behind to the head in front. On each side of the head is situated a large fin (fig. 30, ff), by means of which the animal propels itself through the water. The head is also furnished with indistinct tentacles, and exhibits centrally, on its front margin, the opening of the mouth. The mouth contains a toothed "tongue," essentially similar to that of the Whelk. The mouth opens into a long and slender gullet, which conducts to a stomach, this in turn communicating with a long and slender intestine. There is a well-developed liver, and the intestine terminates in a distinct vent placed on the right side of the neck. The nervous system forms a mass situated below the gullet. There is a heart, consisting of two chambers; but the breathing-organs are quite rudimentary, and can hardly be said to constitute regular gills.

Hyalea tridentata is a native of the Mediterranean and Atlantic Ocean, and is found in the open sea, far from land, darting about by means of the vigorous flapping of the lateral fins. It appears to be of nocturnal habits, and to sink below the surface during the daytime.

RECAPITULATION OF ESSENTIAL CHARACTERS.—The head is furnished with lateral expansions of the skin, or fins, by means of which locomotion is effected. The mouth is furnished with a toothed tongue. The animal lives in the open ocean, near the surface of the water. A non-essential but common character is, that the body is protected by a symmetrical, glassy, semi-transparent shell. These characters distinguish the class of the *Pteropoda* as a whole.

CHAPTER XVIII.

CLASS CEPHALOPODA.

THE next group of animals is a large one, comprising the various kinds of Cuttle-fishes and the Pearly Nautilus, all of which live in the sea. The name of the group is *Cephalopoda* (from the Greek, *kephale*, head, and *podes*, feet), so called because the head is surrounded by a series of "arms," or muscular processes which the animal uses for walking with at the bottom of the sea. As the representative of these animals we shall select the common Calamary (*Loligo vulgaris*) of British seas. This singular creature (fig. 31, A) grows to a length of from a foot and a half to two feet, and is not uncommonly found stranded on the shore after heavy storms. The animal consists, as can readily be seen, of two portions—an anterior or front portion, carrying the eyes, and a posterior or hinder portion, into which the former is loosely fitted in front. The hinder portion is the body proper, and is of a cylindrical or rounded shape, furnished behind with a broad triangular fin on each side. These fins give the hinder end of the body a somewhat lozenge-shaped form, and they enable the animal to swim with great power and rapidity. The whole of the body is enclosed in thick leathery skin, of a bluish colour, and covered with numerous purplish-red specks and blotches. The under surface is of a lighter tint, and the animal can change its colour at will, and can thus adapt itself to the colour of surrounding objects.

The anterior portion of the body carries on its sides a pair of large, conspicuous, globular eyes, and bears in front a circle of muscular processes or "arms." These arms are ten in number, eight of them of equal size, and the remaining two very much longer than the others. The eight short arms (fig. 31, A, *a*) are furnished on their inner surfaces with two rows of little cups, or "suckers," which enable the animal to seize objects firmly, and also to walk about head downwards at the bottom of

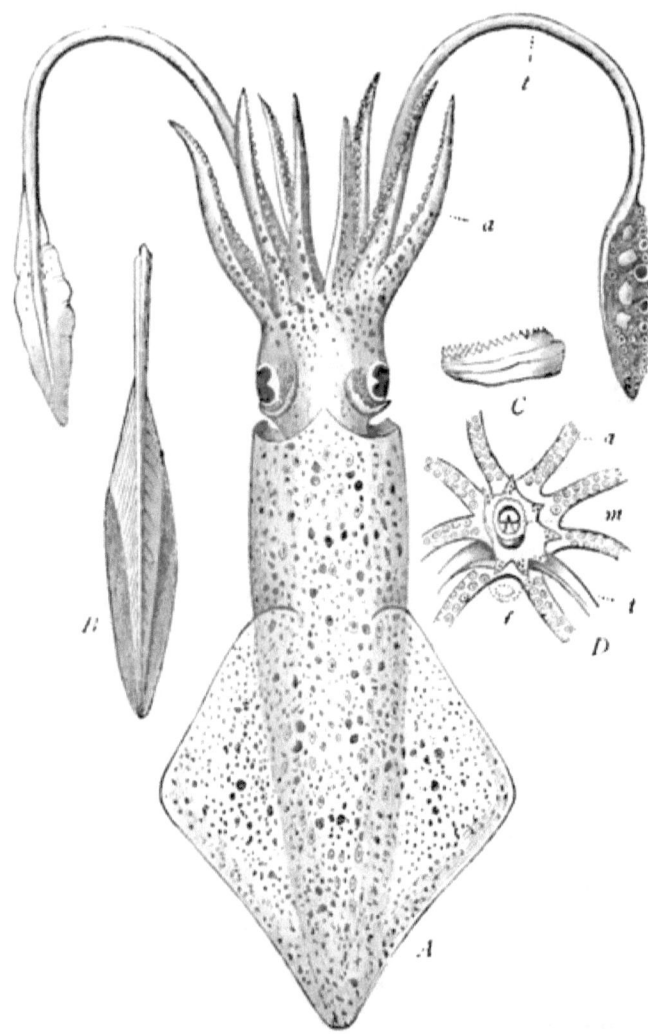

Fig. 31.—A, The common Calamary (*Loligo vulgaris*), reduced in size : *a* One of the ordinary arms ; *t* One of the longer arms or "tentacles." B, Skeleton or "pen" of the same, one fourth natural size (after Woodward). C, Side view of one of the suckers, showing the horny hooks surrounding the margin. D, View of the head from in front, showing the bases of the arms (*a*) and tentacles (*t*), the mouth (*m*), and the funnel (*f*).

the sea. These suckers, numerous as they are, are entirely under the control of the animal; and each is furnished with a ring of horny hooks round its margin (fig. 31, C, so that they constitute collectively a most efficient apparatus for adhesion and for grasping purposes. The two longer arms are known as the "tentacles" (fig. 31, A, t), and they only carry suckers at their extremities, which are expanded and club-shaped.

If we separate the arms a little from one another, so as to expose the front of the head, we see the opening of the mouth, surrounded by the bases of the arms (fig. 31, D, m); and within the mouth is a pair of strong jaws, of a horny consistency, brown with white tips, and very like the beak of a parrot, except that the undermost jaw is the longest.

The only other point in the external anatomy of the animal which needs mention, is a peculiar tube which is seen on the under surface of the head. (This is not visible in fig. 31, A, since this represents the upper surface of the animal, but it is shown in fig. 31, D, f). This tube is called the "funnel," and the animal has the power of ejecting through it a stream or jet of water. By means of this jet, by its reaction on the surrounding water, the animal can propel itself, tail foremost, without the necessity of using its fins. The "funnel" also serves other purposes which will appear hereafter.

Returning now to the mouth, we may briefly examine the internal structure of the animal. The mouth, with its beak-like jaws, opens into a gullet, surrounding which we find a ring of nervous matter (fig. 32, n), which represents the brain of the higher animals, and which is protected by a rudimentary skull. Besides the jaws, the mouth also contains a tongue, the hinder portion of which is covered with spines. The gullet leads into a stomach, from which proceeds an intestine, terminating at the bottom of the "funnel." The funnel, therefore, serves to convey out of the body the undigested portions of the food. There is also a well-developed liver (fig. 32, l), which pours its secretion into the intestine. Placed upon one side of the intestine is a curious organ, which is generally known as the "ink-sac" (fig. 32, *i*). This is a little bag or membranous sac, filled with a jet-black semi-fluid material—the "ink" —which the animal has the power of squirting out at will. The tube which leads from the ink-sac opens at the base of the funnel, and the "ink" can thus be thrown into the water outside. The animal accordingly, when threatened by any danger, emits a jet of this inky fluid, and makes its escape under cover of the cloud which it has thus raised.

The remaining internal organs which mainly concern us, are the heart and gills. The heart receives the pure blood, which has passed through the gills, and distributes it to all parts of the body. The gills are the breathing-organs, and are two in number. They are pyramidal in shape, and the impure or venous blood is submitted in them to the action of the oxygen contained in the water, which is freely admitted to them; and which, after passing over their surface, is again expelled from the funnel.

Lastly, we find that the soft and yielding body of the Calamary is strengthened and supported by a singular

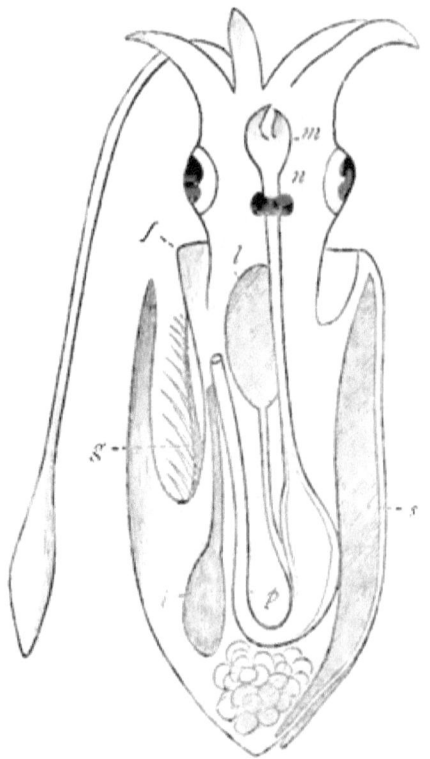

Fig. 32.—Diagram of the internal anatomy of a Cuttle-fish (altered from Huxley). *m* Jaws; *n* Nervous ring surrounding the gullet; *p* Intestine, opening at the base of the funnel (*f*); *i* Ink-sac, also opening at the base of the funnel; *g* Gills; *s* Skeleton.

internal skeleton, which is enclosed within the thick skin of the body. This skeleton (fig. 31, B) has something of the form of a paddle or of a feather, and it is generally known as the "pen." It is of a horny consistence, semi-transparent, and consisting of a central stem with two lateral expansions or "wings." In old specimens there are several of these "pens," packed closely one behind the other.

The Calamary, as before mentioned, lives in the sea, and the present species is generally distributed round the shores of Great Britain. Where abundant, it is used by the fishermen as bait. In the quaint language of Pennant, these animals "inhabit all our seas; are gregarious; swift in their motions; take their prey by means of their arms, and embracing it, bring it to their central mouth. Adhere to rocks, when they wish to be quiescent, by means of the concave discs that are placed along their arms." They appear to live upon shell-fish, and sometimes sea-weed, and they lay their eggs in clusters containing many thousands.

RECAPITULATION OF ESSENTIAL CHARACTERS.—Of the above-mentioned characters which distinguish the Calamary, the following are essential: The body is symmetrically constructed, so that you could divide the animal with a knife into two halves, which would externally be exactly similar to one another. The mouth is placed in the front of the head, and is surrounded by a circle of muscular processes or "arms." The breathing-organs are in the form of gills; and the water which has passed over the gills is expelled from the body by means of a muscular tube or "funnel." By the jet of water thus emitted, the animal can propel itself through the water. The mouth is furnished with beak-like jaws, and also with a tongue, the hinder part of which is provided with bent spines. The nervous system is highly developed, and is partially protected by what may be regarded as a rudimentary skull. These characters distinguish all the animals which are related to the Calamary, and they may therefore be taken as the distinctive characters of the entire class of the *Cephalopoda*.

CHAPTER XIX.

CLASS PISCES.

THE class *Pisces* (Latin, *piscis*, a fish) includes the familiar animals which are properly known as fishes. As the type of this vast and universally distributed group, we may take the common Perch (*Perca fluviatilis*) of the rivers and lakes of Britain and Northern Europe.

The general shape of the body in the Perch (fig. 33) is somewhat spindle-shaped, tapering towards both extremities, and thus adapting the animal for rapid and easy movement in a watery medium. From the tip of

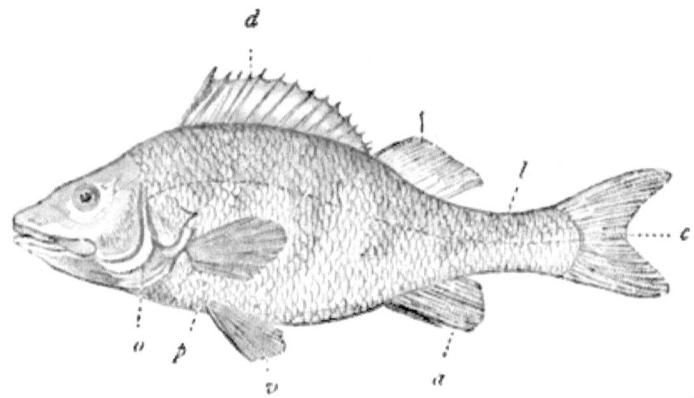

Fig. 33.—The common Perch (*Perca fluviatilis*). *o* Gill-cover, w th the gill-slit behind it; *p* One of the pectoral fins, the left; *v* The left ventral fin; *d* The first dorsal fin; *d'* The second dorsal fin; *c* The caudal fin or tail; *a* The anal fin; *l* Lateral line.

the nose for about a quarter of its whole length the thickness rapidly increases, and then it more slowly diminishes towards the tail for the remaining three-quarters of its length. The body is also considerably compressed or flattened from side to side, and its height is much greater than its width.

The body is covered with a closely-fitting armour of scales developed in the skin. These scales (fig. 36, B) are thin, flexible, horny plates, which overlap one another like the tiles of a house, and which have their hinder edges fringed with a comb-like row of spines. Running from the back of the head to the root of the tail is a peculiar row of scales (fig. 33, *l*), which constitute what is called the "lateral line." Every scale in this line is perforated by a small hole, and communicates internally with a peculiar system of canals. It used to be believed that the slime which covers the body was secreted by these canals, and poured out by the lateral line of scales; but it is more probable that the slime or mucus which covers the body is really the outer layer of the skin.

The fish propels itself through the water by means of the "fins," all of which have the common character of being expansions of the skin in the form of membrane, supported by numerous "rays" or thin bony spines, much as an umbrella is supported by its ribs. Some of the "rays" (fig. 34, A) are undivided bony spines, and are termed "spinous rays;" others are branched towards their ends, and are divided by joints into numerous short pieces, when they acquire the name of "soft rays" (fig. 34, B). Some of the fins have only spinous rays; some have only soft rays; and some possess both kinds of rays combined with one another.

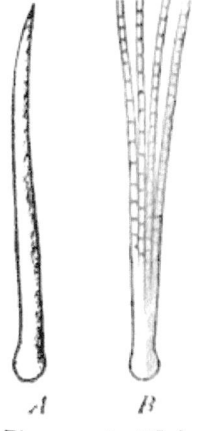

Fig. 34.—A, "Spinous ray" of a fin, in the form of a simple spine. B, "Soft ray" of a fin.

Whilst all the fins agree in consisting of membrane supported by bony "rays," they are nevertheless divided into two very distinct groups, quite different in their real character. Some of the fins, namely, are placed *in pairs* on the sides of the body, and are really the *limbs* of the animal. Others are *unpaired*—that is to say, they are simply placed in the central line of the body, and have no fellows. The "paired" fins are four in number, making two pairs. The front pair (fig. 33, *p*) represents the forelegs, and is termed the "pectoral fins" (Latin, *pectus*, the breast). They are placed on the side of the body, just behind the head. The hinder pair (fig. 33, *v*) represents the hind-legs, and is termed the "ventral fins" (Latin, *venter*, the belly), being placed on the lower surface of the fish, behind and below the pectorals. Both the pectoral and ventral fins of the fish are employed chiefly in steering the animal, and more especially in ascending and descending in the water. All the "rays," lastly, of the pectoral fins are "soft" and jointed; and all those of the ventral fins have the same character, with the exception of the first ray, which is "spinous."

The "unpaired" fins are divided into three sets: one placed above or on the back, one placed at the end of the body, and one placed below. The unpaired fins along the back are two in number (fig. 33, *d, d'*), and are known as the first and second "dorsal fins" (Latin, *dorsum*, the back). The first dorsal is supported by fifteen "spinous" rays, the second dorsal by rays of which all but the first are "soft." The function of the dorsal fins is chiefly to enable the fish to maintain its equilibrium and vertical position in the water. The unpaired fin at the end of the body is commonly called the "tail," but is technically known as the "caudal fin" (Latin, *cauda*, the tail). It is entirely supported by "soft" rays, and is symmetrical in form, consisting of two equal lobes (fig. 33, *c*). The position of the tail is vertical, so that it strikes the water from side to side. Along with the whole flexible hinder end of the body, it constitutes the main organ of progression of the fish, its powerful strokes driving the animal forwards, much as a boat can be propelled by a single oar placed over the stern. Lastly, a single unpaired fin (fig. 33, *a*) is placed upon the lower surface of the body in front of the tail, and is known as the "anal fin" (Latin, *anus*, the vent), because it is placed near the vent. All its rays are "soft," except the first two, which are rigid spines.

The skeleton of the Perch (fig. 35) is very complicated, and can only be briefly alluded to. Its most important

CLASS PISCES. 79

portion is the "backbone," which is seen to form a long axis or central stem, round which the body is symmetrically

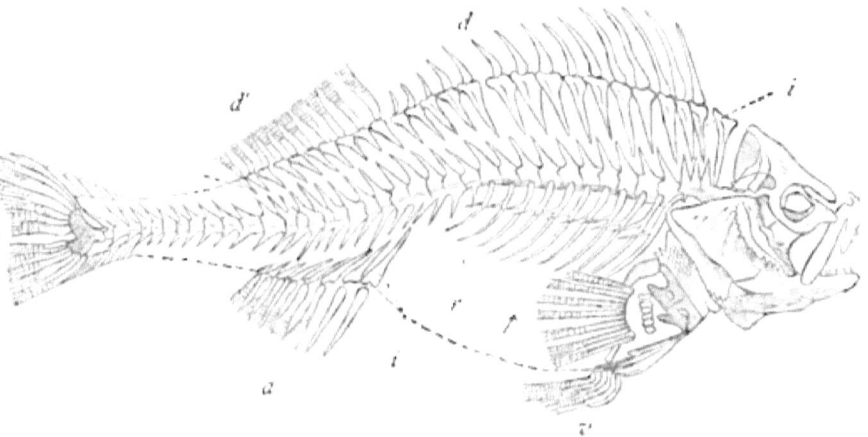

Fig. 35.—Skeleton of the common Perch (*Perca fluviatilis*). *p* Pectoral fin; *v* One of the ventral fins; *a* Anal fin, supported upon interspinous bones (*i*); *c* Caudal fin; *d* First dorsal fin; *d'* Second dorsal fin, both supported upon interspinous bones; *i i* Interspinous bones; *r* Ribs.

built up. The backbone is technically called the "vertebral column," and is composed of a succession of short pieces or segments, each of which is termed a "vertebra" (Latin, *verto*, I turn). Each separate piece or vertebra of the backbone is deeply cupped or hollowed out at each end, and the entire spine is thus rendered extremely flexible. From the possession of this backbone or spinal column, the Perch is said to be a "vertebrate animal." The essential purpose served by the backbone is to protect the very important portion of the nervous system which is known as the "spinal marrow," or "spinal cord." The unpaired fins are all connected with the backbone by means of a series of little bony spines, which are placed in the central line of the body, and are known as the "interspinous bones" (fig. 35, *i*).

At the front end of the backbone is placed the skull, which protects in its interior the brain or central portion of the nervous system.

The mouth in the Perch is placed at the front of the

head, and is armed with small teeth, which have a uniform size, and are bent backwards. The throat is pierced with a series of slits, which, as we shall see, allow the water to reach the gills or breathing-organs. Behind the throat the digestive canal consists of a gullet, stomach, and intestine; and the latter terminates in a distinct vent, placed on the lower surface of the body. There is a large and well-developed liver, and behind the stomach are placed three closed tubes, which are believed to represent another of the digestive glands (the sweetbread or pancreas).

At the sides of the throat, on the under surface of the head, are placed the breathing-organs and heart. The former constitute the "gills," and are adapted solely for breathing air dissolved in water, and not for breathing air directly. Hence the fish dies if removed from its natural element. The gills (fig. 36, A, *b b*) have the form of a series of fringes, supported upon four bony arches (*a*),

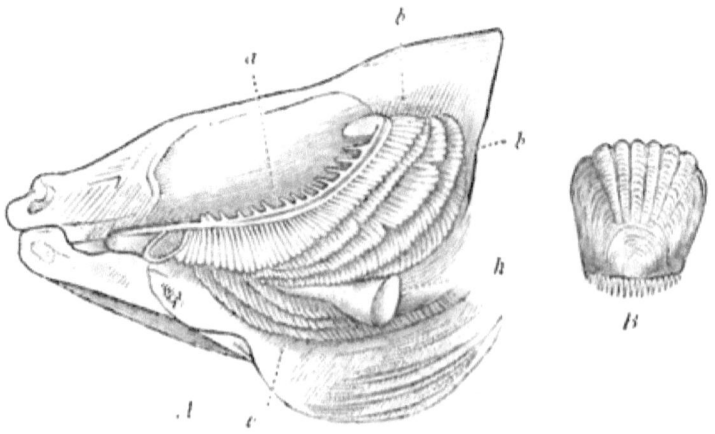

Fig. 36.—A, Gills and heart of the Perch exposed by the removal of the gill-cover on the left side. *a* First of the four bony arches which carry the gills (*b b*); *b'* The lower edges of the gills on the right side; *h* Heart (after Van der Hoeven). B, Scale of the Perch, magnified.

and having a bright red colour from the number of blood-vessels distributed to them. A similar set of gills, supported upon four similar arches, is placed upon the opposite side of the head (*b'*). The heart (*h*) has the function of driving the impure blood to the gills, and it consists of two chambers or cavities only.

The process of respiration or breathing is carried on by the Perch as follows: The gills, as we have just described them, are placed on the sides of the throat; but, instead of being exposed to view, they are concealed and protected on each side by a kind of movable door or flap, composed of a series of flattened bones, and known as the "gill-cover" (fig. 33, *o*). They are also protected below by a membrane, which is supported by slender, curved, bony spines. In this way the gills occupy a sort of chamber on each side of the neck. Internally, each of these chambers opens into the throat by a series of slits, already alluded to; and externally, each opens upon the surface by a vertical slit or fissure, which is placed just behind the gill-cover, upon the side of the neck, and which is known as the gill-slit. When the Perch wishes to breathe, which it does several times in the minute, it opens its mouth and takes in a gulp of water; it then forces the water into the gill-chambers through the slits in the throat; and finally, after the gills have extracted all the oxygen from it, the fish opens the gill-cover and expels the now airless water through the gill-slit.

The Perch also possesses a singular membranous sac or bladder, which is termed the "swim-bladder," and is placed below the spine, behind the head. This is filled with gas, and appears to have the function of enabling the animal to rise and sink in the water at will. It is to be regarded, however, as really corresponding with the *lungs* of the air-breathing vertebrate animals.

The nervous system of the Perch consists mainly of the spinal marrow, protected by the backbone, and the brain, contained within the skull. As organs of sense, the animal possesses two eyes, which are destitute of eyelids; ears which are placed internally, and have no outward opening; and a nose, opening by nostrils on the top of the snout, but having the peculiarity that it does not open into the throat behind.

The Perch is one of the commonest of British and European fishes, exclusively inhabiting fresh water. "It is a gregarious fish, and loves deep holes and gentle streams." It is also exceedingly voracious, and it will live for some hours out of water. It attains a weight of eight or nine pounds; but this is quite exceptional, and a Perch of three pounds weight is unusually large. Its colours, when alive, are very brilliant — green above, passing into yellow below, with dark transverse bands on the sides, and with the unpaired fins bright red.

RECAPITULATION OF ESSENTIAL CHARACTERS. — The animal is a water-breather, and the breathing-organs are in the form of gills. The limbs, when present, are converted into "fins." The heart is (with few exceptions) only two-chambered, and the blood is cold. Nearly universal, though not quite, are the possession of a covering of scales, the existence of a swim-bladder, and the fact that the nose does not open behind into the throat. By these characters the class *Pisces* is distinguished as a whole.

CHAPTER XX.

CLASS AMPHIBIA.

THE *Amphibia* are a class of animals which may be regarded as in many respects intermediate between the true Reptiles and the Fishes, and they derive their name from the fact that they mostly inhabit the water when young, and live upon the land when old, or from the fact that they live indifferently either on the land or in water (Greek, *amphi*, both; *bios*, life). As an example of this class we may take the common Frog of Britain (*Rana temporaria*).

The peculiarities of the Frog will be most clearly understood if we commence with the animal from the egg and trace its development till it assumes its adult characters. The eggs of the Frog are deposited in water in gelatinous masses, and the young Frog on leaving the egg is known as a Tadpole. In this stage of its existence (fig. 37, *a*)

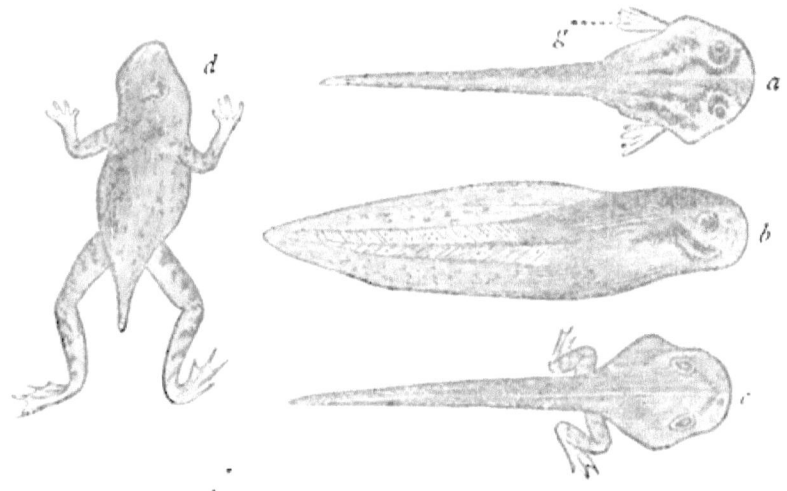

Fig. 37.—Development of the common Frog (*Rana temporaria*); *a* Tadpole, viewed from above, showing the external gills (*g*); *b* Side view of a somewhat older specimen, showing the fish-like tail; *c* Older specimen, in which the hind-legs have appeared; *d* Specimen in which all the limbs are present, but the tail has not been wholly absorbed (after Bell).

the animal resembles a little fish having a large rounded head, a protuberant projecting belly, and a long compressed tail (*b*). Its breathing-organs are at this period genuine gills, two sets of which exist at first. One set of gills is external, and has the form of little filaments placed on the side of the neck (fig. 37, *a*, *g*). These, however, soon disappear, and the animal then breathes by means of a series of internal gills, contained in a kind of chamber, and protected by a flap of skin, which conceals them from view. The water is admitted to these gills by the mouth, and the respiration at this period is carried on essentially in the manner which distinguishes the true Fishes. At this period of its existence, also, the animal not only lives exclusively in the water, but it feeds upon vegetable matters, and it possesses a very long and much-convoluted intestine.

After a while the limbs commence to appear, two pairs of these being proper to the adult animal. The first to become visible externally are the hind-limbs (fig. 37, *c*); whilst the fore-limbs are developed within the chamber which contains the gills, and do not make their appearance till later. The tail, however, still remains; and it is not till the development of the limbs has become considerably advanced that the tail begins gradually to diminish in size and to dwindle away. Finally, the animal takes to the land, and, no longer needing its tail to swim with, this organ is completely absorbed, and the adult thus becomes "tailless."

Coincident, however, with the sprouting forth of the limbs and the diminution of the tail is a more important internal change, by which the animal is enabled to leave the water, and betake itself to the dry land. At first, as we have seen, the animal breathes by gills, and cannot exist out of the water. After a period, however, lungs, adapted for breathing air directly, are developed, and the animal now loses its gills altogether. It now ceases to be an inhabitant of the waters, and, changing its medium, it becomes terrestrial in its habits, though still capable of taking to the water when necessary. The fully-grown Frog, therefore, differs from the young Frog, or Tadpole, in breathing by lungs instead of by gills, in having two pairs of well-developed limbs, and in not possessing a tail. It may be added that the adult Frog lives upon animal matter, slugs, insects, and the like, instead of vegetable food, and that its alimentary canal is much shorter and more simple than that of the young.

The fully-grown Frog (fig. 38) has a short body, a broad head, wide mouth, and long muscular legs. The fore-legs are much shorter than the hind-legs, and terminate in four toes each. The hind-legs, on the other hand, are greatly developed, and terminate in five long toes, which are "webbed," or united by a membrane, so that they constitute very efficient and powerful swimming-organs.

The skin is very soft and moist, quite destitute of scales, and very loosely attached to the muscles beneath.

Fig. 38.—The common Frog (*Rana temporaria*).

The animal can to a great extent breathe by means of the porous skin, and it can thus spend a long period under water without the necessity of using its lungs. In an ordinary way, however, it breathes air directly by means of the lungs, and as it possesses no movable ribs, it obtains the air necessary for breathing by a process resembling swallowing. Hence it is possible to suffocate a Frog, in the summer time, when the animal is active, simply by holding its mouth open, and thus preventing its swallowing the air required for breathing.

The mouth is furnished with a row of minute teeth attached to the upper jaw; but the lower jaw is toothless. The tongue is fleshy, and instead of being rooted behind, it is fixed to the front of the mouth, so that its point is directed down the throat. The animal can protrude it to a considerable length, and uses it for the purpose of catching the small animals upon which it lives. The digestive system is well developed, but calls for no especial mention.

The heart is at first like that of Fishes, consisting of only two chambers, and having only the function of driving the blood to the gills. When the lungs are developed, however, the heart undergoes considerable change, becoming like that of a Reptile. It now consists of three chambers, and is so constructed that the impure blood mixes in it with the pure blood coming from the lungs, and that the body is supplied with this mixture. As a consequence of this, at any rate in part, the Frog is, like the Fishes and Reptiles, *cold-blooded*. In other words, its temperature is very little higher than that of the air or water in which it lives.

The nervous system of the Frog is well developed, con-

Fig. 39.—Skeleton of the common Frog (*Rana temporaria*); *d* Vertebræ of the back, with long side-processes.

sisting of a brain and spinal marrow, and of the nerves connected with these. There are also two large eyes, furnished with movable eyelids, and two ears, which can be detected externally.

The skeleton of the Frog (fig. 39) is of a much higher type than that of a fish; but there are only three points which need be particularly pointed out. Firstly, the broad and flat skull is jointed to the backbone by two distinct joints. Secondly, the joints or vertebræ of the back (*d*) carry long side-processes, but the ribs are quite rudimentary and merely consist of gristle. Thirdly, the limbs are not in the least like the "fins" of fishes, but they exhibit the same bones as are present in the limbs of the higher animals.

The common Frog is of a reddish-brown or yellowish-brown colour, spotted with black; but it has, to a certain extent, the power of altering its colour according to the intensity of the light to which it may be exposed. It feeds upon small slugs, worms, and insects, and is in turn largely eaten by various animals, especially by Owls. It is an excellent swimmer, and mostly passes the winter under water, in a torpid condition. Owing to the length of its hind-legs, it is also a capital jumper; and it can produce the peculiar sound which is known as "croaking."

RECAPITULATION OF ESSENTIAL CHARACTERS. — The animal when young is a water-breather, and is provided with gills. In its adult state it breathes by means of lungs (with or without the original gills). Both pairs of limbs are usually present, and they do not present the form of "fins," possessing, on the contrary, the same bones as in higher animals. The blood is cold. The skull is jointed to the backbone by two joints, and the nose opens behind into the throat. The heart is three-chambered in the adult, but two-chambered in the young. The skin is almost universally soft and destitute of scales. These characters distinguish the class *Amphibia* as a whole.

CHAPTER XXI.

CLASS REPTILIA.

The class *Reptilia* (Latin, *repto*, I crawl) comprises all the animals which are commonly called "Reptiles," except the Amphibians, which would also be placed here by popular consent, but which we have seen really to constitute a distinct division. The living reptiles fall into the four very distinct groups of the tortoises and turtles, the snakes, the lizards, and the crocodiles and alligators. It is difficult to choose any single type which adequately represents the entire class, and which is at the same time

Fig. 40.—The Rattlesnake (*Crotalus horridus*).

readily obtainable for examination. We may, however, select one of the venomous serpents, such as the Rattlesnake, most of the remarks which will be made about this form applying with equal truth to the common Viper of Britain.

The Rattlesnake (*Crotalus horridus*) has entirely the form which we generally understand by the term "snake-like," having a long, cylindrical body, tapering towards the tail (fig. 40). The body is completely destitute of limbs, and is covered with an armour formed of small overlapping horny scales. Some of the scales which cover the head are of larger size than the others, and there is also a row of oblong shields which are continued along the whole lower surface of the animal from the head to the end of the tail. The tail terminates in a singular organ known as the "rattle," which the animal shakes when alarmed or intending to bite. The rattle is really a mere appendage of the skin, and is composed of a number of horny pyramidal joints, loosely united together. The animal, lastly, sheds its skin periodically.

The head is somewhat triangular, broadest behind (fig. 40), and supported upon a comparatively slender neck. The eyes have the peculiarity that they are not furnished with movable eyelids, but are covered by a transparent continuation of the skin. This gives the animal a peculiar fixed and stony stare. Between the eye and the nostril on each side is also to be noticed a deep depression or pit.

At the front of the head is placed the mouth, within which we shall find some of the most characteristic structures in the anatomy of the Rattlesnake. The tongue is forked, capable of being protruded from the mouth at will, and, when protruded, maintained in rapid vibration. The tongue, however, in spite of appearances, is a perfectly harmless organ, and the offensive weapons of the snake are to be found in the teeth (fig. 41). The two halves of the lower jaw are very loosely united together in front, so as to allow of their free separation, and they carry each a series of pointed conical teeth, directed backwards, and amalgamated with the substance of the jaw itself (*l*). Behind, each half of the lower jaw is united with the skull by a peculiar movable bone (*q*), which is known as the "quadrate bone" (Latin, *quadratus*, four-sided). This bone is directed backwards behind the base of the skull,

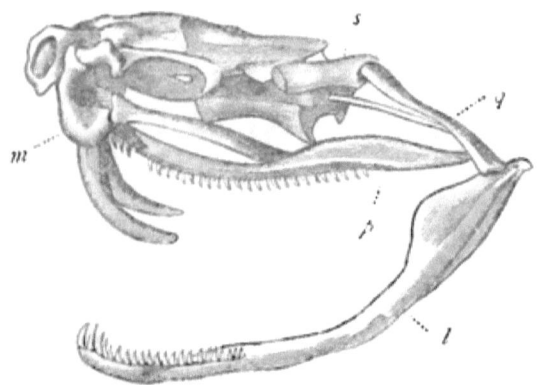

Fig. 41.—Skull of the Rattlesnake (after Dumeril and Bibron). *l* One-half of the lower jaw, united to the skull by the quadrate bone (*q*) ; *m* Upper jaw carrying the poison-fang ; *p* Series of teeth upon the palate.

and owing both to this circumstance and to its mobility, the snake can open its mouth to an extraordinary width, and swallow morsels of comparatively immense size.

Along the palate above, on each side, is also a row of small conical recurved teeth (*p*); but the upper jaws offer the most striking peculiarities. The upper jaw (fig. 41, *m*) is a short movable bone, so jointed to the skull that it can be raised and depressed at will. Each upper jaw carries a great curved tooth, firmly amalgamated with the bone of the jaw, and termed the "poison-fang." The poison-fangs are not only of much greater size than the other teeth, but they are provided with a canal, which opens at the point of the fang in a minute aperture. The canal in the fang communicates above (fig. 42) with the tube or duct leading from a singular organ known as the "poison-gland." This organ (fig. 42, *a*) is a gland placed beneath and behind the eye, and secreting the peculiar fluid which renders the bite of the snake fatal. When the snake wishes to kill an animal for food, or when attacked by an enemy, it employs these formidable weapons in the following manner : The great poison-fangs which were previously directed backwards along the roof of the mouth, are now erected, by the elevation of the

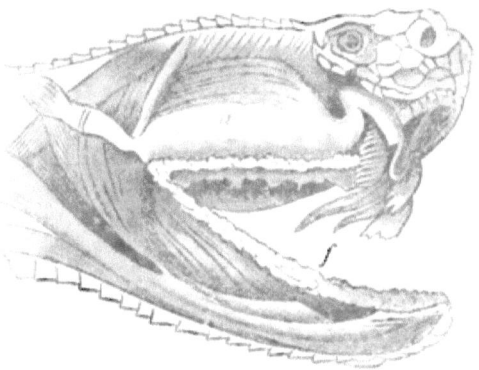

Fig. 42.—The head of the Rattlesnake, dissected to show the poison-gland (*e*) and poison-fangs (*f*). After Duvernoy.

upper jaw-bone, to which they are attached. The fangs are then plunged with great force into the body of the animal which the snake wishes to kill. In performing this act, the muscles which cover the poison-gland compress it strongly, so that as the poison-fangs enter the flesh, a drop of the poisonous fluid is driven into the wound through the little perforation in the point of the poison-fang. The poison itself is a colourless, greasy fluid, and it produces death, at any rate in the majority of cases, with a rapidity proportioned to the size of the wounded animal.

The mouth opens into a comparatively simple digestive tube, which terminates in a vent placed on the under surface of the tail. The breathing-organs are in the form of voluminous lungs, of which one is rudimentary. It is to be remarked, therefore, that the animal is strictly an air-breather, and that at no time of its life does it possess gills, or organs adapted for breathing air dissolved in water. The heart consists of three chambers, and is so constructed that the impure (venous) blood, mixes to a greater or less extent with the pure blood which has come from the lungs; and the body is then supplied with this mixture. As one result of this, the temperature of the body is low, and the animal is "cold-blooded." The nervous system, lastly, consists of a brain and spinal cord, with the nerves which these give off.

It still remains to say a few words about the skeleton of the snake, and especially as to the peculiarities which are connected with its mode of locomotion. The brain is protected within a bony skull, and the spinal cord is protected by a very long and extremely flexible backbone, with which the skull unites by a single joint. There are no traces of limbs, and the snake progresses by gliding upon its belly, walking in reality upon the ends of its ribs. In accordance with this, the ribs are exceedingly numerous, and instead of a number of them being united to a breast-bone in front, this latter bone is absent, and the ribs are simply connected with the horny shields which cover the belly. The snakes creeps along the ground, therefore, by the movements of the numerous ribs, which it employs in progression somewhat in the same way that a centipede uses its legs.

The common Rattlesnake of North America is usually about three or four feet in length, sometimes more. Its general colour is brownish-yellow, with two rows of partially united brown blotches of an irregularly lozenge-shaped figure, but the tail is black. The "rattle" is light brown in colour, and increases in length with the years of the animal, though not receiving an additional joint per annum, as is often stated. Full-grown specimens ordinarily possess a rattle of from sixteen to twenty-four joints. The snake shakes the rattle when alarmed or about to strike, but what the function of this singular organ may be, must, in spite of recent speculations, be regarded as still unknown. The Rattlesnake lives upon small animals, such as hares or squirrels, and birds, which it kills by its poisonous bite, and then swallows whole, the teeth permitting neither division of the food nor mastication. It is a sluggish animal, which remains torpid during the winter, and is most active and most poisonous in the hottest weather.

RECAPITULATION OF ESSENTIAL CHARACTERS.—The animal is an air-breather, and never possesses gills at any time of its life. The two sides of the heart communicate with one another (in most cases), and the body is always supplied with a mixture of pure and impure blood. The blood is cold. The skin usually develops horny scales. The lower jaw is jointed to the skull by means of a "quadrate bone," and the skull is united with the backbone by means of a single joint. The condition of the limbs varies, but in no case are more than two pairs present. These characters distinguish the class *Reptilia* as a whole.

CHAPTER XXII.

CLASS AVES.

THE class *Aves* (Latin, *avis*, a bird) includes only those most familiar and beautiful of animals, the Birds. Instead of finding any difficulty in selecting an example which is both common and at the same time exhibits the leading peculiarities of the class, it would not be an easy matter to choose a common bird in which the more important characters of the entire division are not present. We shall therefore select as our type the domestic Goose, or rather the Grey Lag Goose (*Anser ferus*) from which the domestic breed is descended.

Like all birds, the Goose is a genuine biped, supporting its body, in standing or walking, exclusively upon its hind-legs. As in the great majority of birds, also, the fore-limbs are converted into wings, and are employed in supporting the body of the animal in the air, or, in other words, in flight. The fore-limbs are thus useless for purposes of grasping, and all acts of this nature are performed by the beak—that is to say, by the jaws; though some birds can likewise employ the hind-feet in seizing objects.

The entire body of the Goose is covered with a close covering of those peculiar appendages of the skin which are

Fig. 43 —The Grey Lag Goose (*Anser ferus*). After Yarrell.

known as feathers. The lower portion of the legs and the beak are unfeathered, but the head, neck, and body are protected by a dense plumage, which serves two purposes. In the first place, since it conducts heat very imperfectly, the plumage serves to retain and conserve the heat generated within the body of the animal. The Goose, therefore, though spending a portion of its time in a medium so cold as water, whilst it is a "hot-blooded" animal, is nevertheless able to keep its temperature up to as high as about 100° or a little over. In the second place, the plumage is kept oiled by the oily secretion of certain glands near the tail, and the bird is thus enabled to enter the water without getting wet.

The feathers carried by the wing are longer than those which cover the body, and are the organs which propel

CLASS AVES.   95

the bird through the air. Those carried by the hand are longer than those supported by the fore-arm, and the whole can be made to beat the air by the downward stroke of the wing. The feathers of the tail, also, though of no great length in the Goose, are longer than those which cover the body, and can be spread out like a fan, so as to serve as a rudder and guide the bird in its course through the air.

Whilst the wings or fore-limbs are used in flight, the hind-limbs or legs are used both in walking upon the land and in swimming in the water. The legs are placed so far back (fig. 43) that the animal cannot walk lightly or gracefully, but on the contrary has a waddling and awkward gait upon the land. This same circumstance, however, enables the feet to act very efficiently as oars or paddles,

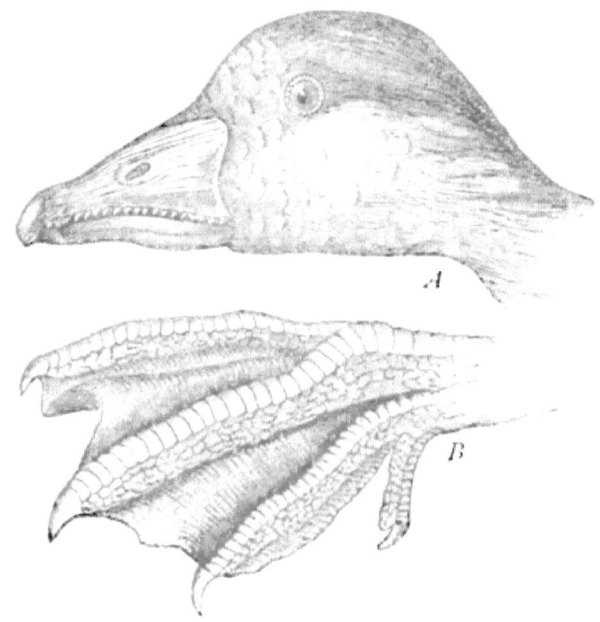

Fig. 44.—A, Head of the Grey Lag Goose ; B, Foot of the domestic Goose.

when the animal is in the water, and this use of them is greatly facilitated by the peculiar form of the foot. The foot (fig. 44, B) consists of four toes, of which three are turned forwards and one backwards. The toe which is turned backwards is really the innermost toe of the foot (in man the "great toe"), and it is not only much smaller than the others, but is raised above the ground, so as to be comparatively useless. The three toes which are turned forwards are all united with one another by skin or membrane, so that the foot becomes "webbed," and thus strikes the water like a broad paddle. The feet and lower portions of the legs are also unfeathered, and are protected by horny plates.

The head carries the eyes, the openings of the ears (concealed beneath the feathers), and the bill. The eyes are furnished with an upper and lower eyelid, and also with a third eyelid, which can be drawn across the eye from the inside, so as to protect the organ from an excess of light.

The bill is composed of the upper and lower jaws, encased in horn, and quite destitute of teeth. The lower jaw is united to the skull, much as we saw in the serpent, by means of the peculiar bone, which is termed the "quadrate bone." On the top of the upper half of the bill, at about its middle, we see the openings of the nostrils, in the form of two oval slits. The bill is broad and soft, only horny towards its margins, which are furnished with numerous parallel transverse plates. Within the bill is a large tongue, much more fleshy than is ordinarily the case amongst birds.

As regards the digestive system, the most noticeable point is that a portion of the stomach is converted into an extremely strong muscular organ, which is termed the "gizzard," and in which the food is ground down and rendered fit for absorption into the blood. As the bird, therefore, does not possess teeth, it may really be said to chew its food in the gizzard.

The blood is hot, and the heart is a four-chambered organ, the two sides of which do not communicate in any way. There is, therefore, no such mixture of the pure and impure blood as takes place in Reptiles.

The Goose, though to a great extent organised for an

aquatic life, is nevertheless strictly an air-breather; and the breathing-organs are in the form of spongy lungs placed within the cavity of the chest. The air, also, is admitted from the lungs to a number of air receptacles or chambers placed in different parts of the body. Further, the air is admitted to the interior of a considerable number of the bones, which are hollow, and filled with air in place of marrow. By means of these arrangements, the body of the animal is considerably lightened, and the blood is likewise more completely exposed to the action of the air than could otherwise have been the case.

The nervous system consists mainly of a brain, protected within the skull, and a spinal marrow, protected within the backbone.

The peculiarities of the skeleton of the Goose are so numerous that nothing can be done here beyond merely pointing out one or two of the more important ones. One of the most striking of these is the form of the breastbone, which has the shape of a broad plate carrying a great central ridge or keel of bone. The object of this is to offer a greatly extended surface for the attachment of the very powerful muscles required for the movement of the wings. Again, the neck is very long and flexible, and is composed of a number of joints, so as to allow the bird to apply its bill to all parts of the body; an arrangement rendered necessary by the fact that the limbs cannot be used in grasping or laying hold of objects. The last joint of the tail, again, is a large ploughshare-shaped bone, which is set on nearly at right angles to the rest of the backbone, and which serves for the elevation and depression of the feathers of the tail. Finally, the skull is united with the backbone by a single joint or pivot.

Lastly, the Geese, like all other birds, produce their young in the form of eggs, and, like most birds, sit upon these eggs so as to hatch them by the warmth of their own bodies. When hatched, the young birds are quite active and can run about and look for food, whilst their bodies are protected against the cold by an ample covering of soft down.

The Grey Lag or Grey-Legged Goose seems to be unquestionably the parent of the domestic breeds of Geese.

It was formerly very common in England, but is now very rare. According to Gould, it "is known to inhabit all the extensive marshy districts throughout the temperate portions of Europe generally; its range northwards not extending further than the fifty-third degree of latitude, while southwards it extends to the northern portions of Africa, eastwardly to Persia, and, we believe, is generally dispersed over Asia Minor." The Grey Lag Goose lives upon grass, water-plants, and seeds, and abides mainly in fens and marshy places. It not only swims and flies well, but its walk, owing to the height of its legs, is not so clumsy as is the case with many water-fowl.

RECAPITULATION OF ESSENTIAL CHARACTERS.—The animal is an air-breather, and at no time of its life possesses gills. The breathing-organs are in the form of lungs, which communicate with air-receptacles, and (almost invariably) with the interior of a greater or less number of the bones. The blood is hot, and the heart four-chambered. The skull is jointed to the backbone by a single joint, and the lower jaw is joined to the skull by means of a "quadrate bone." The skin is provided with the peculiar appendages which are known as feathers. The fore-limbs (when not rudimentary) are organised for flight, and constitute wings. The hind-limbs, in a state of nature, are never provided with more than four toes. The mouth is destitute of teeth, and the jaws are sheathed in horn. The animal brings forth its young in the form of eggs. These characters distinguish the class *Aves* as a whole.

CHAPTER XXIII.

CLASS MAMMALIA.

THE class *Mammalia* (Latin, *mamma*, the breast) includes the ordinary Quadrupeds, all of which suckle their young for a longer or shorter time after birth. As an example of this class we may select the domestic Dog (*Canis familiaris*).

The Dog is a "quadruped" in the strict sense of this term, since it possesses four limbs, two fore-legs and two hind-legs, and all of these are employed in supporting the weight of the body. Unlike many quadrupeds, however, the Dog does not place the soles of his feet upon the ground, but on the contrary walks upon the tips of the toes, thus obtaining his light and springy gait. The soles of the feet, therefore (fig. 46, *m m*), are raised above the ground, and are covered with hair. The fore-feet have five toes each, the hind-feet (as a regular thing) only four toes each; and all the toes are furnished with strong claws. The claws of the Dog, however, unlike those of the cat, are always exposed to view, and cannot be withdrawn into sheaths.

In place of the feathers which are so distinctive of Birds, the skin of the Dog is covered with the peculiar appendages known as "hairs;" and these appendages are characteristic of the Mammals, in which they are never and at all times of life altogether wanting.

The body of the Dog is supported internally by a complicated skeleton (fig. 46), or series of bony structures, which protects the internal organs from injury, and serves for the attachment of the muscles by which all the movements of the animal are effected. In briefly examining some of the more important points connected with this skeleton, it may be considered as consisting of the skull, the backbone and its appendages, and the bones of the limbs. The skull (fig. 45) is a strong bony box or case within which that most important organ, the brain, is protected.

Besides the true brain-case, however, the skull also

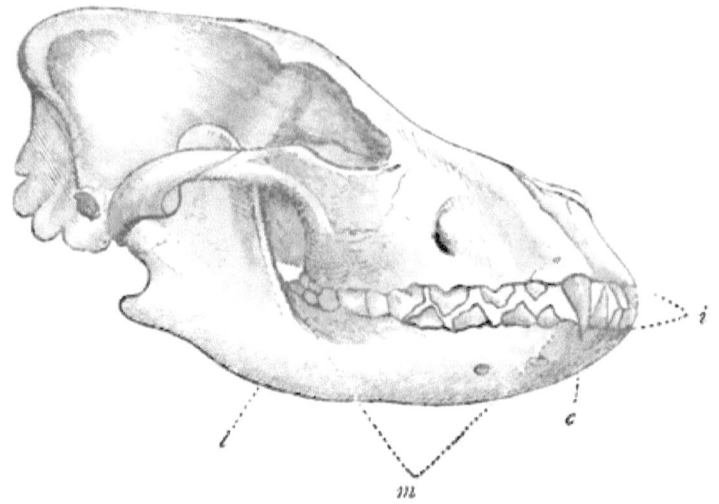

Fig. 45.—Skull of the Sheep-Dog. *l* Lower jaw ; *m* Back teeth or molars ; *c* Eye-teeth or canines ; *i* Front teeth or incisors.

consists of the bones of the face, and likewise serves for the protection of the organs of vision and hearing. The bones of the face are much lengthened out, so as to form a regular muzzle, at the end of which the mouth opens. The lower jaw (*l*) is composed of two halves, each consisting of a single piece, and it is jointed to the skull behind, directly, and not by means of any "quadrate bone," such as exists in Reptiles and Birds. Both jaws are armed with teeth, which are sunk in distinct sockets in the bony substance of the jaw. In the front of each jaw are placed six teeth, the so-called front teeth or "incisors" (Latin, *incido*, I cut into), which have sharp cutting-edges, and are used in dividing the food (fig. 45, *i*). On each side of these are situated the so-called eye-teeth or "canine teeth" (Latin, *canis*, a dog), two in each jaw, and therefore four in number altogether. The canine teeth (fig. 45, *c*) are very long and pointed, and their special function is to act as weapons of offence and defence, and also to kill the animals upon which the Dog feeds. Outside the canine teeth, again, are situated the teeth which are known commonly as the back teeth or "grinders," and which are technically called the "molars" (Latin, *mola*, a mill). The molar teeth (fig. 45, *m*) of the Dog are twelve in number in the upper jaw (six on each side), and fourteen in number in the lower jaw (seven on each side). They can, however, only very partially be called "grinders," with any propriety, since all but the last two on each side of each jaw have sharp edges adapted for cutting and dividing flesh.

The skull is united to the backbone behind by two joints. The first part of the backbone (fig. 46, *n*), imme-

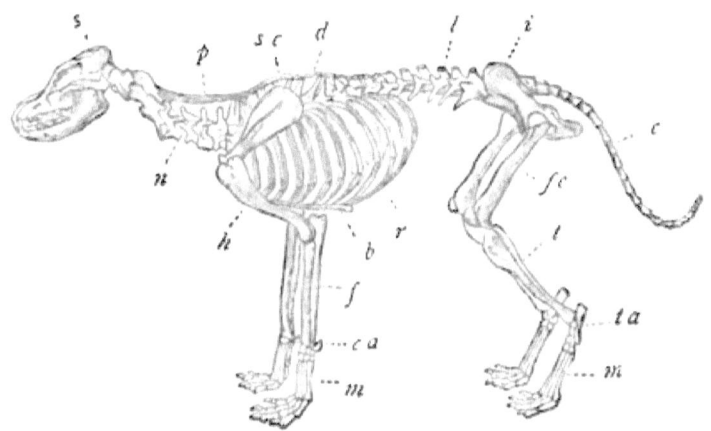

Fig. 46.—Skeleton of the Dog (after Youatt). *s* Skull; *n* Neck; *d* Back; *l* Loins; *c* Tail; *sc* Shoulder-blade; *h* Bone of the upper arm; *f* Two bones of the fore-arm; *ca* Bones of the wrist; *m* Palm of the hand; *i* Haunch-bone; *fe* Thigh-bone; *t* Bones of the shank or shin; *ta* Bones of the ankle; *m* Sole of the foot; *r* Ribs; *b* Breast-bone; *p* Ligament supporting the head.

diately behind the head, constitutes the neck, and is composed of seven joints. The next portion of the backbone constitutes the back properly so called, and the different joints or "vertebræ" here carry the ribs (*r*). The ribs form the walls of the chest, protecting the heart and lungs, and they are connected in front with the bone known as the "breast-bone" (*b*). Behind the true back comes that portion of the backbone which is known as the "loins" (*l*), which is followed by a portion which is enclosed between the two haunch-bones (*i*). Finally, there is the long and slender portion of the backbone which constitutes the "tail" (*c*).

The fore-limb is connected with the trunk chiefly by means of the "shoulder-blade" (*sc*), followed below by the strong and single bone of the upper arm (*h*). This is succeeded by two bones which constitute the fore-arm (*f*), and these carry a number of little bones which make up the wrist-joint (*ca*). The palm of the hand (*m*), as already remarked, is carried off the ground, and the hand finally terminates in five fingers, or rather toes.

The hind-limb is connected with the trunk by means of the haunch-bone (*i*), which carries the upper bone of the leg or "thigh-bone" (*fe*). Below this are the two bones which constitute the shin (*t*), and these carry the small bones of the ankle (*ta*). The instep and sole of the foot (*m*) are raised completely off the ground, and the foot terminates in four toes.

As regards the digestive system of the Dog, the jaws, as we have seen, are armed with teeth; but these teeth are intended for cutting and not for bruising or grinding down the food. The Dog, therefore, cannot be said to chew or "masticate" its food, but it swallows its food in separate mouthfuls which it simply tears or cuts off with its sharp teeth. Salivary glands (Latin, *saliva*, spittle) pour their secretion into the mouth, thus facilitating swallowing; and the tongue, unlike that of the cat, is quite smooth instead of being prickly. The mouth opens into a gullet, the gullet into a capacious stomach, and the stomach into an intestine, but the last is comparatively short as compared with the intestine of vegetable-eating quadrupeds. The intestine also receives the secretions of two large digestive glands—namely, the liver and the "sweetbread" or "pancreas" (Greek, *pan*, all; *kreas*, flesh)—both of which produce fluids which exercise an important influence on the food.

The food is converted by the action of the digestive organs into a white, milky fluid, which is then absorbed from the intestine by a series of special vessels, and poured into the blood. The blood is thus composed of the elaborated products of digestion, and, like that of all quadrupeds, it is warm, having an average temperature of $98°$. The blood is distributed to all parts of the body by means of a system of closed tubes—the blood-vessels,—the propelling force being derived from a central contractile organ—the heart. The heart is four-chambered, and is so constructed that the impure blood which has circulated through the body, does not mix with the pure blood coming from the lungs. On the contrary, the blood which has become impure by contact with the tissues is always submitted to the action of the oxygen of the air in the breathing-organs, before it is again allowed to be driven through the body.

The breathing-organs of the Dog are in the form of two lungs placed within the cavity of the chest, the animal being strictly an air-breather, and never having gills at any time of its life. The air is admitted to the lungs by a tube which opens into the throat, and is known as the "windpipe," and the lungs never communicate with air-receptacles as in Birds.

The main masses of the nervous system of the Dog consist of the brain, protected within the skull, and the spinal cord, protected by the backbone. The senses are well developed, with the exception of that of touch, in the delicacy of which the Dog is far exceeded by man.

The innumerable breeds of the domestic Dog, in spite of the enormous differences between them, seem to be descended from no more than three or four wild types, of which the wolf and jackal are the most important. Some of the most striking varieties of the Dog, such as the greyhound, appear to be exceedingly ancient, and it is not possible to say positively from what wild stock they have descended. The chief characters which distinguish the more typical breeds of Dog, such as the sheep-dog, from the wolf, are the possession by the former of a recurved tail and circular pupil of the eye. According to Cuvier, the Dog is "the most complete and most useful conquest that man has made. The whole species has become our property; each individual is entirely devoted to his master, adopts his manners, distinguishes and defends his property, and remains attached to him even unto death; and all this springs not from mere necessity, nor from constraint, but simply from gratitude and true friendship. The swiftness, the strength, and the highly developed power of smelling of the dog, have made him a powerful ally of man against the other animals, and were perhaps necessary to the establishment of society. It is the only animal that has followed man all over the earth."

RECAPITULATION OF ESSENTIAL CHARACTERS.—The animal is an air-breather, and at no time of its life possesses gills. The breathing-organs are in the form of lungs, which do not communicate with air-receptacles or with the interior of the bones. The blood is hot, and the heart four-chambered. The skull is united to the backbone by a double joint, and the lower jaw is joined to the skull directly, and not by means of a "quadrate bone." The skin is provided with the peculiar appendages which are known as hairs. The animal brings forth its young alive, and the young are nourished for a longer or shorter time by the mother by means of a special secretion—the milk—secreted by means of special glands—the mammary glands.

CHAPTER XXIV.

THE SUB-KINGDOMS.

IT remains very briefly to consider how the classes of which we have examined representative types, can be arranged into larger divisions. The possibility of such an arrangement depends upon the fact that certain classes possess characters in common, which characters are not shared by other classes. By attending to these general characters, which belong to several groups, we can arrange the classes of the Animal Kingdom into six larger divisions, which are called "sub-kingdoms" (Latin, *sub*, under), because they are the primary sections into which the whole kingdom may be broken up. The following gives briefly the characters of each sub-kingdom and the classes contained in it—two or three small classes, not here considered, being omitted :—

SUB-KINGDOM I. PROTOZOA.

The sub-kingdom *Protozoa* (Greek, *protos*, first ; *zoön*, animal) derives its name from the fact that it comprises the lowest forms of animal life known to the naturalist. It includes the two classes of the *Rhizopoda* and *Infusoria*, which exhibit the following common characters :—The body exhibits no distinct rings or segments, and the walls of the body do not enclose any definite cavity which could be properly termed a "body-cavity." They have either no proper digestive system, or, at most, a mouth and short gullet. There is no nervous system, and the only representative of a blood-circulatory system is to be doubtfully found in one or more little contractile chambers, which are not universally present. Examples of the sub-kingdom are the *Amœba*, the Sponges, and *Paramœcium*.

SUB-KINGDOM II. CŒLENTERATA.

The name of *Cœlenterata* or "hollow-entrailed" animals is applied to this sub-kingdom because the creatures included in it have a body-cavity which communicates with the world outside through the mouth (Greek, *koilos*, hollow ; *enteron*, intestine). The sub-kingdom includes the two classes of the *Hydrozoa* and *Actinozoa*, and is defined by the following characters :—The body is composed of two distinct layers, an outer and an inner, and the latter of these includes a distinct cavity, which is properly called a "body-cavity." There is a distinct mouth, which opens either directly into the body-cavity, or into a distinct stomach. The stomach, however, when present, opens below into the body-cavity, so that the latter in all cases communicates directly with the outer world through the mouth. The integument is provided with the singular microscopic organs of offence and defence which are known as "thread-cells" or "nettle-cells." There is no distinct blood-system or heart; and the nervous system is either absent or quite rudimentary. The body does not exhibit any distinct segmentation, but some of its parts are always arranged in a more or less star-like or "radiate" manner. Examples of the sub-kingdom are the Fresh-water Polypes (*Hydra*), the Sea-anemones (*Actinia*), the Sea-firs, Sea-jellies, Corals, &c.

SUB-KINGDOM III. ANNULOIDA.

The name *Annuloida* (Latin, *annulus*, a ring; Greek, *eidos*, form) is given to this sub-kingdom because of the resemblance of some of the animals which it comprises to the true *Annulosa* or Ringed animals, such as the Worms. This sub-kingdom includes the two classes of the *Echinodermata* and *Scolecida*, and is characterised by the following peculiarities:—The body-cavity is completely shut off from the digestive cavity (when this is present), so that the body-cavity in no case communicates with the outer world through the mouth. There is a distinct nervous system, and sometimes a blood-system. In all, there is a peculiar system of "water-vessels," consisting of canals which are distributed through the body, and in most cases receive water from the outside through an aperture in the walls of the body. The fully-grown animal is sometimes more or less star-like or "radiate" in the arrangement of its parts, sometimes more or less elongated and worm-like. In no case, however, does the body consist of a regular series of segments placed one behind the other, nor are their limbs disposed symmetrically on the two sides of the animal. Examples of the sub-kingdom are the Star-fishes, Sea-urchins, Tape-worms, Round-worms, Wheel-animalcules, &c.

SUB-KINGDOM IV. ANNULOSA.

The name *Annulosa* is derived from the fact that the animals included in this sub-kingdom possess bodies which are composed of a succession of distinct segments or "rings" (Latin, *annulus*, a ring). The sub-kingdom includes the classes of the *Annelida, Crustacea, Arachnida, Myriapoda,* and *Insecta*, which agree with one another in the following characters :—The body is composed of successive rings or segments placed one behind the other in a longitudinal series. The body-cavity never communicates with the exterior through the mouth. There is mostly a well-developed digestive system, which is always shut off from the body-cavity. The nervous system consists of a double chain of little nervous masses placed along the "ventral" or lower surface of the body and united by longitudinal cords. The limbs (when present) are turned towards that side of the body upon which the nervous system is situated, and they are mostly arranged in symmetrical pairs on the two sides of the body. Examples of the sub-kingdom are the Leeches, Earth-worms, and Sea-worms; the Lobsters, Shrimps, and Crabs; the Spiders and Scorpions; the Centipedes and Millipedes; and the true insects (Beetles, Butterflies, Dragon-flies, House-flies, and the like).

SUB-KINGDOM V. MOLLUSCA.

The name of this sub-kingdom is derived from the fact that the animals comprised in it have soft bodies (Latin, *mollis*, soft); and they are commonly called "Shell-fish," because the soft body is generally protected by a hard covering or shell. This sub-kingdom includes the classes of the *Polyzoa, Tunicata, Brachiopoda, Lamellibranchiata, Gasteropoda, Pteropoda,* and *Cephalopoda,* which agree in the possession of the following characters :—The animal is soft-bodied, and does not exhibit either distinct segmentation, or any star-like arrangement of its parts. Usually, but not universally, the soft body is protected by a hard outer covering or shell. The body-cavity does not communicate with the outer world through the mouth, and there is always a well-developed digestive system which in no case communicates directly with the body-cavity. The nervous system consists of a single principal mass or of scattered masses not arranged in a chain. There is usually a distinct heart, and generally distinct breathing-organs, but both may be absent. Examples of this sub-kingdom are the Sea-mats, Sea-squirts, Lamp-shells, Bivalve Molluscs (Oysters, Mussels, and the like), Univalve Molluscs (Snails, Whelks, and the like), Cuttle-fishes, &c.

SUB-KINGDOM VI. VERTEBRATA.

This sub-kingdom derives its name from the fact that the animals comprised in it have a more or less perfectly developed backbone (Latin, *vertebra,* a bone or joint of the spine or backbone). The sub-kingdom includes the classes of the *Pisces* (Fishes), *Amphibia, Reptilia, Aves* (Birds), and *Mammalia* (Quadrupeds), all of which agree with one another in the following characters :—The body is composed of a number of definite segments placed one behind the other in a longitudinal series. The body-cavity never communicates with the outer world through the mouth, and there is always a well-developed alimentary canal which never opens into the body-cavity. A distinct blood-system and distinct breathing-organs are invariably present. The nervous system is placed in a cavity distinct from the general body-cavity, and is situated along the back of the animal. With one exception, the nervous system consists of a distinct brain and spinal cord, and the latter is protected from injury by a more or less well-developed backbone or "vertebral column." The limbs (when present) are never more than four in number, and are turned away from that side of the body upon which the main masses of the nervous system are situated. Examples of the sub-kingdom are the various kinds of Fishes, the Newts, Frogs, and Toads; the Tortoises, Snakes, Lizards, and Crocodiles; the Birds; and the numerous species of true Quadrupeds or Mammals.

INDEX.

ABDOMEN, of Lobster, 29; of Spider, 37; of Centipede, 39; of Insect, 42.
Actinia, 14; form of, 14; tentacles of, 16; nettle-cells of, 16; digestive system of, 16; habits of, 16.
Actinozoa, 14; characters of, 17.
Adductor muscles of *Mya*, 58.
Æshna (*see* Dragon-fly).
Amœba, 6; form of, 6; feeding of, 7; locomotion of, 7; contractile chamber of, 8; habitat of, 8.
Amphibia, 82; characters of, 87.
Anal fin of Perch, 78.
Annelida, 23; characters of, 27.
Annuloida, characters of, 106.
Annulosa, characters of, 107.
Anser (*see* Goose).
Antennæ, of Lobster, 31; of Spider, 37; of Centipede, 39; of Insect, 43.
Appendages of Lobster, 31, 32.
Arachnida, 35; characters of, 38.
Arms, of Star-fish, 17; of Terebratula, 55; of *Loligo*, 71.
Articulate animals, 33.
Ascidia, form of, 51; coverings of, 52; breathing-organs of, 53; digestive system of, 53; heart of, 53; nervous system of, 53; habits of, 53.
Ascidians (see *Tunicata*).
Aves, 93; characters of, 98.

BEAK, of the shell of *Terebratula*, 54; of the shell of *Mya*, 62; of the Goose, 96.
Bivalve shells, 56; of *Terebratula*, 54; of *Mya*, 58.
Brachiopoda, 54; characters of, 56.
Breathing-organs, of Star-fish, 20; of Leech, 26; of Lobster, 33; of Spider, 37; of Centipede, 41; of Insect, 45; of *Flustra*, 50; of *Ascidia*, 53; of *Mya*, 60; of Whelk, 67; of Cuttle-fish, 74; of Perch, 81; of Frog, 86; of Rattlesnake, 91; of Goose, 97; of Dog, 103.
Buccinum (*see* Whelk).

CALAMARY, form of, 71; arms of, 71; jaws of, 73; funnel of, 73; digestive system of, 73; nervous system of, 73; ink-bag of, 74; heart of, 74; breathing-organs of, 74; pen of, 75; habits of, 75.
Canine teeth of Dog, 101.
Canis (*see* Dog).
Carriage-spring apparatus of *Terebratula*, 55.
Caudal fin of Perch, 78.
Centipede, 39; head of, 39; foot-jaws of, 39; body-rings and legs of, 40; eyes of, 39; digestive system of, 41; heart of, 41; breathing-organs of, 41; nervous system of, 41; habits of, 41.
Cephalopoda, 71; characters of, 75.

INDEX.

Cilia, of *Paramœcium*, 9; of Wheel-animalcule, 21.
Cœlenterata, characters of, 105.
Crotalus (*see* Rattlesnake).
Crustacea, 27; characters of, 34.
Cuttle-fishes, 71.

DIGESTIVE system, of *Paramœcium*, 10; of *Hydra*, 13; of *Actinia*, 16; of Star-fish, 20; of Wheel-animalcule, 22; of Leech, 26; of Lobster, 33; of Spider, 38; of Centipede, 41; of Dragon-fly, 44; of *Flustra*, 49; of Ascidia, 53; of *Terebratula*, 56; of *Mya*, 60; of Whelk, 67; of *Hyalea*, 70; of Calamary, 73; of Perch, 80; of Frog, 85; of Rattlesnake, 91; of Goose, 96; of Dog, 102.
Dog, form of, 99; hairs of, 99; skull of, 100; jaws of, 100; teeth of, 100; backbone of, 101; limbs of, 102; digestive system of, 102; blood of, 103; heart of, 103; breathing-organs of, 103; nervous system of, 103.
Dorsal fin of Perch, 78.
Dragon-fly, head of, 42; eyes of, 42; jaws of, 43; thorax of, 44; wings of, 44; abdomen of, 44; digestive system of, 44; heart of, 45; breathing-organs of, 45; nervous system of, 45; metamorphosis of, 45, 46; habits of, 46.

Echinodermata, 17; characters of, 20.
Eosphora (*see* Wheel-animalcule).
Eyes, of *Actinia*, 16; of Star-fish, 20; of Wheel-Animalcule, 23; of Leech, 26; of Lobster, 31; of Spider, 36; of Centipede, 39; of Dragon-fly, 42; of Ascidia, 52; of Whelk, 65; of Calamary, 71; of Perch, 81; of Frog, 86; of Rattlesnake, 89; of Goose, 96.

FEATHERS of Goose, 94.

Feet of Star-fish, 19.
Fins of Perch, 77.
Flustra, form of, 47; cells of, 47-49; tentacles of, 49; digestive system of, 49; nervous system of, 50; habitat of, 50.
Foot, of *Mya*, 57; of Whelk, 65.
Foot-jaws, of Lobster, 31; of Centipede, 39.
Fresh-water Polype (*see Hydra*).
Frog, development of, 83, 84; form of, 84; skin of, 85; mouth and tongue of, 85; heart of, 86; breathing-organs of, 86; nervous system and sense-organs of, 86; skeleton of, 87; habits of, 87.

GAPER (*see Mya*).
Gasteropoda, 64; characters of, 69.
Gill-cover of Perch, 81.
Gills, of Lobster, 33; of *Mya*, 59; of Whelk, 67; of Calamary, 74; of Perch, 80; of Tadpole, 86.
Gizzard of Goose, 96.
Goose, form of, 93; feathers of, 94; hind limb and foot of, 95; eyes of, 96; beak of, 96; digestive system of, 96; heart of, 96; lungs of, 97; nervous system of, 97; skeleton of, 97; habits of, 97, 98.

HEAD, of Lobster, 29; of Spider, 35; of Centipede, 39; of Dragon-fly, 42.
Head-shield of Lobster, 31.
Heart, of Lobster, 33; of Spider, 38; of Centipede, 41; of Dragon-fly, 45; of *Ascidia*, 53; of *Mya*, 60; of Whelk, 67; of Calamary, 74; of Perch, 81; of Frog, 86; of Rattlesnake, 91; of Goose, 96; of Dog, 103.
Homarus (*see* Lobster).
Horn-wrack (*see Flustra*).
Hyalea, shell of, 69; fins of, 70; digestive system of, 70; heart and breathing-organs of, 70;

nervous system of, 70; habitat of, 70.
Hydra, form of, 12; tentacles of, 12; thread-cells of, 13; budding of, 13; habits of, 14.
Hydrozoa, 11; characters of, 14.

INCISOR teeth of Dog, 100.
Infusoria, 9; why so called, 9; characters of, 11.
Ink-sac of Calamary, 74.
Insecta, 42; characters of, 46.
Interspinous bones of Perch, 79.

JAWS, of Leech, 25; of Lobster, 31; of Spider, 37; of Centipede, 39; of Dragon-fly, 43; of Calamary, 73; of Perch, 79; of Frog, 85; of Rattlesnake, 89; of Goose, 96; of Dog, 100.

Lamellibranchiata, 56; characters of, 64.
Lamp-shells, 54.
Lateral line of Perch, 77.
Leech, form of, 24; suckers of, 24; locomotion of, 24; jaws of, 25; digestive system of, 26; breathing-organs of, 26; nervous system of, 26.
Ligament of the shell of *Mya*, 62.
Lithobius (see Centipede).
Lobster, 27; structure of the body of, 29; segments and appendages of, 29-33; digestive system of, 33; gills of, 33, 34; heart of, 33; nervous system of, 34; habits of, 34.
Loligo (see Calamary).

MAMMALIA, 99; characters of, 104.
Mammary glands, 104.
Mantle, of *Terebratula*, 54; of *Mya*, 57; of *Hyalea*, 70.
Metamorphosis of Dragon-fly, 45, 46.
Molar teeth of Dog, 101.

Mollusca, characters of, 107.
Mya, 57; mantle of, 57, 58; adductor muscles of, 58; gills of, 59; digestive system of, 60; respiratory process of, 60; shell of, 62, 63; habits of, 63.
Myriapoda, 39; characters of, 41.

NERVOUS system, of Star-fish, 20; of Wheel-animalcule, 22; of Leech, 26; of Lobster, 34; of Spider, 38; of Centipede, 41; of Dragon-fly, 45; of *Flustra*, 50; of *Ascidia*, 53; of *Terebratula*, 56; of *Mya*, 60; of Whelk, 67; of Calamary, 73; of Perch, 81; of Frog, 86; of Rattlesnake, 92; of Goose, 97; of Dog, 103.
Nervures of the wings of Dragon-fly, 44.
Nettle-cells, of *Hydra*, 13; of *Actinia*, 16.
Nipping-claws of Lobster, 31.

OPERCULUM of Whelk, 66.

PAIRED fins of Perch, 77.
Paramœcium, 9; cilia of, 9; digestive system of, 10; contractile chamber of, 10.
Pectoral fins of Perch, 78.
Pen of Calamary, 75.
Perca (see Perch).
Perch, form of, 76; scales of, 77; lateral line of, 77; fins of, 77, 78; skeleton of, 79; digestive system of, 80; gills of, 80; heart of, 81; respiratory process of, 81; swim-bladder of, 81; nervous system and sense-organs of, 81; habits of, 82.
Phallusia (see *Ascidia*).
Pisces, 76; characters of, 82.
Poison-fangs, of Spider, 37; of Centipede, 39; of Rattlesnake, 90.
Poison-gland of Rattlesnake, 90.
Polyzoa, 47; characters of, 50.
Proteus-Animalcule (see *Amœba*).
Protozoa, characters of, 105.
Pteropoda, 69; characters of, 70.

INDEX.

QUADRATE bone, of Rattlesnake, 89; of Goose, 96.

RADIATE type of structure, 14.
Rana (see Frog).
Rattlesnake, form of, 89; scales of, 89; rattle of, 89; eyes of, 89; tongue of, 89; jaws of, 89; teeth of, 90; poison-gland of, 90; digestive system of, 91; breathing-organs of, 91; heart of, 91; nervous system of, 92; skeleton of, 92; habits of, 92.
Rays of the fins of Perch, 77.
Reptilia, 88; characters of, 93.
Respiratory sacs, of Leech, 26; of Spider, 37; of *Ascidia*, 53.
Rhizopoda, 6; characters of, 8.
Ringed Worms (see *Annelida*)

Sanguisuga (see Leech).
Scales, of Perch, 77; of Rattlesnake, 89.
Scolecida, 21; characters of, 23.
Sea-anemone (see *Actinia*).
Sea-mat (see *Flustra*).
Sea-squirts (see *Ascidia*).
Shell, of *Terebratula*, 54; of *Mya*, 62, 63; of Whelk, 66; of *Hyalea*, 69.
Siphons, of *Mya*, 59; of Whelk, 66.
Skeleton, of Perch, 79; of Frog, 87; of Rattlesnake, 92; of Goose, 97; of Dog, 101.
Soft rays of fins of Perch, 78.
Spider, head and thorax of, 35; legs of, 35; eyes of, 36; jaws of, 37; abdomen of, 37; spinnerets of, 37; breathing-organs of, 37; digestive system of, 38; heart of, 38; nervous system of, 38; habits of, 38.
Spinnerets of Spider, 37.
Spinous rays of fins of Perch, 78.
Star-fish, form of, 17; skin of, 18; feet and water-vessels of, 19, 20; digestive system of, 20; nervous system of, 20; respiration of, 20; habits of, 20.

Sub-kingdoms, 104.
Suckers of Calamary, 72.
Swim-bladder of Perch, 81.
Swimmerets of Lobster, 32.

TADPOLE of Frog, 83.
Tail, of Lobster, 29; of Perch, 78.
Teeth, of Perch, 80; of Frog, 85; of Rattlesnake, 90; of Dog, 100.
Tegenaria (see Spider).
Tentacles, of *Hydra*, 12; of *Actinia*, 14; of *Flustra*, 49; of Calamary, 73.
Terebratula, 54; shell of, 54; arms of, 55; digestive system of, 56; nervous system of, 56; habits of, 56.
Thorax, of Lobster, 29; of Spider, 35; of Dragon-fly, 44.
Thread-cells, of *Hydra*, 13; of *Actinia*, 16.
Tongue, of Whelk, 67; of *Hyalea*, 70; of Calamary, 73; of Frog, 85; of Rattlesnake, 89.
Tunicata, 51; characters of, 53.
Tunics of *Ascidia*, 52.

UNIVALVES, 64.
Unpaired fins of Perch, 78.
Uraster (see Star-fish).

VENTRAL fins of Perch, 78.
Vertebra, 79.
Vertebrata, characters of, 108.

Waldheimia (see *Terebratula*).
Water-vessels, of Star-fish, 20; of Wheel-animalcule, 22.
Wheel-animalcule, 21; ciliated wheel of, 21; digestive system of, 22; water-vessels of, 22; nervous system of, 22.
Whelk, 64; foot of, 65; siphon of, 66; shell of, 66; tongue of, 67; digestive system of, 67; heart and breathing-organs of, 67; habits of, 68.
Wings of Dragon-fly, 44.

www.ingramcontent.com/pod-product-compliance
Lightning Source LLC
Chambersburg PA
CBHW020138170426
43199CB00010B/801